风景创意压花艺术

[美] 朱少珊 著 \ 世界压花协会推荐参考书

中国林业出版社
China Forestry Publishing House

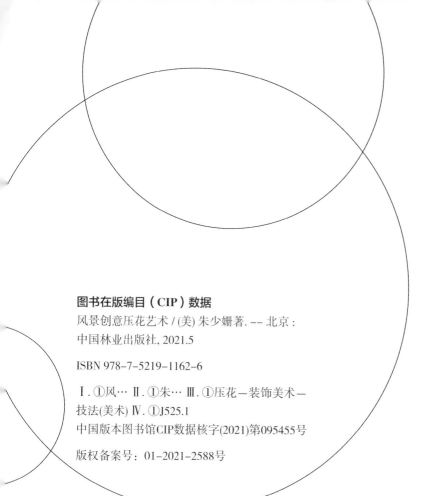

图书在版编目（CIP）数据

风景创意压花艺术 / (美) 朱少姗著. -- 北京：中国林业出版社, 2021.5

ISBN 978-7-5219-1162-6

Ⅰ.①风… Ⅱ.①朱… Ⅲ.①压花—装饰美术—技法(美术) Ⅳ.①J525.1

中国版本图书馆CIP数据核字(2021)第095455号

版权备案号：01-2021-2588号

策划编辑：印芳

责任编辑：印芳　邹爱

出　版	中国林业出版社（100009 北京西城区刘海胡同 7 号）
电　话	010-83143565
发　行	中国林业出版社
印　刷	河北京平诚乾印刷有限公司
版　次	2021 年 6 月第 1 版
印　次	2021 年 6 月第 1 次印刷
开　本	710 毫米 × 1000 毫米　1/16
印　张	8.5
字　数	140 千字
定　价	59.00 元

前言
Preface

  自从我的初级、中级和高级压花艺术三本书出版了之后，一直试图写关于创意压花艺术和关于从高级迈向大师级压花艺术方法和技巧的书。对我而言，这是一项非常艰巨的任务。我希望这套书《风景创意压花艺术》《静物创意压花艺术》能往压花的艺术方面引导我的读者，而不仅仅是简单的作品制作方法。

  压花艺术有什么用？ 艺术让我思考，让我去探索和想象。使我欣赏到美学价值。约翰·拉伯克（John Lubbock）写道："毫无疑问，艺术是人类幸福中最纯粹，最高的元素之一。它通过眼睛训练大脑，并且通过大脑训练眼睛。如同太阳赋予花色彩，艺术赋予生活色彩。"我希望压花艺术能为每个人的生活增色。

  在我写作期间，我先生庆承侃为我写了首诗。在此分享一下。他的鼓励，是我坚持写出这套书的一大动力。并且感谢他为我的压花作品提出自己的看法和意见。

      纤手玉指理奇葩，
      花草枝叶尽入画。
      当年黛玉若识君，
      红楼梦里无葬花。

  书里通过不同的范例课程和图片，引导读者试验压花艺术不同的创意、各种方法和技巧并掌握它们。因此，每个人都可以使用这些方法和技巧来创作出属于自己的作品。

After publishing three beginner to advanced pressed flower art books, I had been trying to write a book on an even higher level of methods and techniques. It has been a very difficult task for me. I want this book to guide my readers on the art side of pressed flowers. This is not a simple how-to book.

What is pressed flower art for? Art makes me think, enables me to explore, and allows me to imagine. It makes me appreciate the value of aethetics. John Lubbock wrote, "Art is unquestionably one of the purest and highest elements in human happiness. It trains the mind through the eye, and the eye through the mind. As the sun colors flowers, so does art color life.". I hope that pressed flower art colors everyone's lives.

This book contains many step-by-step projects and sample pictures. My goal is to guide readers to experiment with all kinds of creative ideas, to try out the methods and techniques, and to master them. Thus, everyone can create their own artwork with these methods and techniques.

I want to thank my husband for the encouragement and support during my writing.

2021.04

目录
Contents

前言 Preface	003
压花艺术创意和技巧 Pressed Flower Art Ideas and Techniques	008
压花艺术欣赏 Pressed Flower Art Appreciation	011
关于压花艺术设计 About Pressed Flower Art Design	016
关注中心（或重点）Center of interest (or Emphasis)	016
平衡 Balance	018
统一 Unity	019
韵律 Rhythm	019
设计分析 Analysis of Designs	020
光 Light	022
剪影和影子 Silhouette and Shadow	024
黑色压花材料 Black Pressed Materials	025
剪影设计 Silhouette Design	026
影子 Shadow	026
背景 Background	027
制作画 Making the picture	027
制作背景 Making the background	028
倒影 Reflection	029
倒影定律 Law of Reflection	029
水面的特性 Water Conditions	030
作品材料 Materials	031
实物和倒影压花材料 Real and Reflection Pressed Materials	031
背景 Background	032
制作森林和倒影 Making the Forest and Reflection	033
制作雾 Making the Mist	033
制作船和人 Making the Boat and People	034
制作前景植物 Making the Foreground Foliage	035
负空间—落日与水 Negative Space- Sunset and Water	036
材料 Materials	038
绘制图样 Sketch the Picture	039

制作作品 Making the Picture 040

色彩和谐
Color Harmony 044
互补色 Complementary colors 046
分割互补色 Split Complementary 047
类似色 Analogous 048
三向轴 Triadic 050
四角色 Tetradic 052
自由配色 Your Own Color Pallet 053

透视
Perspective 054
材料 Materials 056
 一点透视构图 One-point Perspective Drawing 056
 矮树丛的底色 Color Base for Bushes 057
 制作街道 Making the street 057
 制作人行道 Making the Side Walk 057
 制作矮树丛 Making the bushes 058
 制作树顶底色 Making the Base Color of the Tree Tops 058
 制作树干和树枝 Making the Tree Trunk and Branches 059
 制作蓝花楹的色彩变化和光点
 Making the Color Variation and Light Spots of Jacaranda 060
 加入小鹿增加画作的有趣度
 Adding The Young Deer for Visual Interest 061

风景画设计
Landscape Design 064
主题 Theme 066
三分法则 Rule of Third 066
确定焦点 Decide Focal Point 067
构图 Composition 067
建筑元素技巧 Archtectural Element Techniques 070
 材料 Materials 070
 制作步骤 Procedure 071
 房屋 Houses 072
 石墙 Stone Wall 074
 背景 Background 076
 中景制作 Middle Ground 077
 近景制作 Foreground 078
基本风景画元素技巧 Basic Landscape Element Techniques 080
 材料 Materials 081

计划 Planning — 082
- 天空 Sky — 084
- 山 Mountains — 085
- 悬崖 Cliff — 086
- 水和岸背景 Water and Shoreline Background — 087
- 岩石 Rocks — 088
- 树木 Trees — 089
- 草原高地 Grassy Highland — 096
- 野花 Wild Flowers — 096
- 溪流 Stream — 097
- 平静的水面 Calm Water — 098

压花瀑布 Pressed Flower Waterfall — 099
- 材料 Materials — 100
- 压制鳄梨壳窍门 Tips on Pressing Avocado Shell — 100
- 压制花叶青木窍门 Tips on Pressing Aucuba — 102
- 创作理念 Concept — 102
- 关于光线和环境 About Light and Environment — 103
- 制作背景 Making the background — 103
- 制作山丘石头 Making the Rocky Hills — 105
- 制作瀑布 Making the Waterfall — 110
- 制作水雾 Making the Mist — 113
- 制作水中的石头 Making the Rocks in the Water — 116
- 石头上的苔藓 Moss Growth on Stone — 117
- 山谷边缘的树 Tree Branches on the Edge of Canyon — 118
- 定稿细节调整 Final Adjustments — 119
- 装裱注意 Note on Framing — 119

海浪 Wave — 120
- 简介 Introduction — 120
- 材料 Materials — 120
- 海的色彩 Colors of The Sea — 121
- 天空的色彩 Colors of The Sky — 122
- 设计理念 Design Concept — 122
- 构图和底稿 Sketch and Drawing — 123
- 制作天空 Making the Sky — 125
- 制作山丘 Making the Mountains — 126
- 制作大海色彩 Making the Color Base for the Sea — 127
- 制作海滩 Make the Beach — 128
- 制作浅水 Make the Shallow Water — 129
- 设定海浪 Define Waves — 129
- 制作涌浪 Making the Swell Wave — 130
- 制作翻滚浪 Making swirl wave — 130
- 增加几行浪 Add Rows of Waves — 131
- 增强阴影 Shadow Strengthening — 132
- 飞溅 Splash — 133
- 泡沫 Foam — 133
- 涟漪 Ripples — 134
- 远处的波浪 Distance Waves — 134

后记 Postface — 136

01 压花艺术创意和技巧

Pressed Flower Art Ideas and Techniques

我在初、中、高级压花艺术书籍中教授了压花方法、基本美术构图理论和制作各种物件的步骤。这本书志在唤起你灵感的同时让你掌握更多压花艺术的制作理论、方法与技巧。这本书集中讲述风景类的压花设计。带你在创作美丽的压花艺术大道上前进。

I have taught pressing methods, the basics of art composition, and processes of making many items in my beginner, intermediate, and advanced pressed flower art books. This book is for you to be inspired by creative ideas and to master more pressed flower art theory, methods, and techniques. This book concentrates on landscape designs. You are on your way to creating beautiful pressed flower art!

　　我喜欢旅行，到不同的地方旅行会给我带来不同的生活经历。无论是工作还是休闲，旅行都丰富了我的生活。当然，我也曾经遭受误班机，汽车在马路中间熄了火，或是错过了重要的会议这些糟心事。但是那些去过的地方以及与我见过的人之美好回忆成就了我的压花艺术。不仅丰富了我的生活而且也因艺术丰富了其他人的生活。

I love to travel. Traveling to different places brings me different life experiences. Whether it was for work or for leisure, traveling enhances my life. Sure, I have had experiences of missing my flight, of my car dying in the middle of the road, and have had missed important meetings. However, the fond memories about the places I went and the people I met have inspired my pressed flower art that not only enriches my life but the lives of others as well.

　　您在自己的生活中是否有过一些瞬间，让您感动得无法用言语来形容？您是否有创作某些东西的冲动，但又不知道该如何开始或怎么制作？还是您压了很多花但不知道该如何巧妙地利用？ 这本书将为您提供思路，带您探索并教您如何利用压花获得艺术效果的方法和技巧。在这里，花朵和叶子不再仅仅是花和叶，您可以自由地使用它们作为美术媒介创作出属于你的艺术品。

Have you had the urge to create something but don't know where to start or how to do it? Or you simply have pressed a lot but don't know what to do with your flowers? This book will give you ideas to explore and will teach you methods and techniques on achieving artistic effect with pressed flowers. Here, flowers and leaves are not just in their pure form anymore. You can use flowers and leaves as an art medium to freely create your own unique piece of art.

压花艺术创意和技巧　Pressed Flower Art Ideas and Techniques

压花艺术欣赏
Pressed Flower Art Appreciation

想看见花的人总会看得见。

——马蒂斯

There are always flowers for those who want to see them.

——Henri Matisse

光是画中最重要的一员

——莫奈

Light is the most important person in the picture.

——Claude Monet

压花艺术创意和技巧 Pressed Flower Art Ideas and Techniques

在你破碎的废墟上，
这藤蔓仍然会蓬勃生长，
稀有，
新鲜，
如同旧日的芬芳。
爱不会崩溃。

——艾莉娜·法琼

Upon your shattered ruins where

This vine will flourish still,

as rare,

As fresh,

as fragrant as of old.

Love will not crumble.

——Eleanor Farjeon

我在森林里迷失但找到了灵魂。

——约翰·缪尔

And into the forest I go to lose my mind and find my soul.

——John Muir

关于压花艺术设计
About Pressed Flower Art Design

  以下是有关风景画设计的"规则"。规则并非牢不可破，有时打破它们会给我们带来震撼的效果。但是，在刚开始设计时，遵循这些规则通常会给我们带来愉悦的心情和谐的画面。

The following give we some "rules" on landscaping designs. Rules are not unbreakable. Sometimes, departing from these rules will give we a dramatic effect. However, following these rules will generally give we a pleasant and harmonious arrangement - especially when we first start out in design.

关注中心（或重点）Center of interest (or Emphasis)

  每幅画都需要有一个关注中心，使观众更好地理解我们的画，在这里，要向观众展示作品的主要思想。准备设计作品时，必须先牢记要描绘的特定主题、想法或对象。当作品中没有任何可吸引观众到特定区域或物体的关注中心时，观众的眼睛会在整个场景中徘徊。关注中心可以是单个物体，也可以是一组物体，能让观众把注意力集中在某个特定的区域上。

This is where we will present the main idea of our picture to viewers. Every picture needs to have a center of interest in order to provide viewers a focal point to understand our picture.

The specific topic, idea, or object to be portrayed must be set in our mind as we prepare to make a picture. When there is nothing in the picture to attract attention to a particular area or object, the eyes wander throughout the scene. The center of interest may be a single object or numerous ones arranged so attention is directed to one definite area.

位置 Placement
  要在哪里展示我们的关注中心？最简单的方法是三分法。
  如果在画面垂直和水平绘制两条线，将图像区域垂直和水平均等地分为3个部分，则相交点周围就是您要放置关注中心的位置。

Where do we present your center of interest? The simplest way is by the rule of thirds.

If we draw two lines vertically and horizontally, diving the image area into three parts equally both vertically and horizontally, the intersection points are where we would want to place our center of interest.

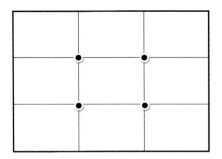

要：

在关注中心使用最明亮的颜色。

在作品的4个角落使用不太强烈的颜色。

不要：

请勿将关注中心置于作品的"正中心"。

地平线不要安排在画的正中间。

Do's:

Use the brightest colors in the center of interest.

Use less intense colors at the corners of the picture.

Do not's:

Do not place the center of interest in the "center" of the picture.

Do not have horizon in the middle of the picture.

简单化 Simplicity

简单是获得优质画作的关键。作品越简单直接，意思就能传达得越清晰，越强。当我们讨论简单化时，有几件事要考虑。首先，选择一个适合自己的主题。例如，与其让整个画面都布满能迷惑观众的元素，不如缩小区域，将聚焦点在一些重要元素上。确保背景中没有任何东西可以分散观看者对画作重点的注意力。同样，请检查前景中是否有任何东西挡住人的目光走进作品。

Simplicity is the key to most good artwork. The simpler and more direct a picture is, the clearer and stronger is the resulting statement. There are several things to be considered when we discuss simplicity. First, select a subject that lends itself to a simple arrangement; for example, instead of a picture of an entire area that would confuse the viewer, focus on some important elements and do a smaller area. Be sure there is nothing in the background to distract the viewer's attention from the main point of the picture. Likewise, check to see there is nothing in the foreground that blocks the entrance of the viewer's eye into the picture.

最后一点是讲简单的故事。需要确保图片中有足够的元素可以传达一个想法。尽管每幅作品都由许多小区域和相关元素组成，但是没有什么比作品的关注中心吸引更多观看者的注意力了。关注中心传递是您制作画作的首要原因。因此，所有其他元素应仅以支持和强调关注中心为目的。请勿使场景混乱，或使用偏离作品关注中心的意思的元素和线条。

A last point of simplicity: tell only one story. Ensure there is only enough material in the picture to convey a single idea. Although each picture is composed of numerous small parts and contributing elements, none should attract more of the viewer's attention than the primary object of the picture. The primary object is the reason the picture is being made in the first place; therefore, all other elements should merely support and emphasize the main object. Do not allow the scene to be cluttered with confusing elements and lines that detract from the primary point of the picture.

平衡 Balance

画作需要平衡。这并不意味着我们需要在各处散布相同强度的颜色。较深色的颜色比较浅的颜色比重要重。因此，一小块强烈的色彩能平衡大面积的较浅色。

平衡并不意味着所有对象的大小均相等。靠近图片中心的较大物体会平衡靠近侧面的较小物体。

您可能希望拥有一个主要的对象，然后再拥有一组较小的对象（如同在不对称设计一样）。

Picture needs to be balanced. It does not mean that we need to spread the same intensity of colors everywhere. Intense color weighs more than lighter colors. Therefore, a small area of intense color balances out a larger area of lighter colors.

Balancing does not mean all the objects are equal in sizes. Bigger objects closer to the center of picture would balance out smaller objects closer to the sides.

You would want to have one dominate object and then other smaller object as in an asymmetrical design.

统一 Unity

好的设计可以使作品中的所有元素都融为一体。因此要仅包括属于场景的事物。

A good design makes all the elements in the picture cohesive. Only include things that belong to the scene.

韵律 Rhythm

韵律使画面生动,它使用重复的图案(有或没有变化)来吸引读者的眼睛随着画移动。

Rhythm means using repetitive patterns (with or without variations) that help viewers' eyes moving along the picture. Rhythm brings life to the picture.

设计分析
Analysis of Designs

艺术家有责任带领观众观赏您的作品。人们倾向于像我们看书一样看画，从上到下，从左到右。下面看一下不同的设计如何引领观众观赏我们的作品。

It is the responsibility of the artist to lead viewers to read your artwork. People tend to look at pictures just like how we read books – from top to bottom and left to right.

Take a look below how different designs allow us to lead viewers along our artwork.

观众通常会根据小径、道路、水路等等来观看我们的作品。

Viewers usually follow a path, road, waterway, etc....to look our work.

画面的左边大块陆地造成画面失去平衡。右边水域太大太开阔使观众目光一下子便到了画面之外。

This picture is out of balance on the bottom left due to the heavy land mass. The lake opening wide on the right leads the viewer out of the picture.

这次平衡稍微好点，但还不够。

This is a little better balanced but not enough.

画面达到平衡但是海岸线比较无趣。水域通道从左到右。所有的元素都太平整。

Picture is balanced but the coastal line is boring. The water way spans across the entire picture. Everything is too flat.

画面达到平衡,海岸线因为有弧度而比较有趣。水域通道比较平,从左到右。

Picture is balanced and the coastal line is much more interesting with slopes. Water opening from left to right.

画面平衡,水域通道是有趣的S型。几条通道都指向我们的关注中心——帆船。

Picture is balanced and the lake is very interesting with the S line. Several path of lines all leads to the center of interest – the boats.

不要把移动的元素放到要离开画面的位置、我们想把观众的目光留在画面上。

Try not placing the traveling objects going away from the picture. We want the viewer's eyes focusing in the picture.

最终我们选择这个设计。

Finally, we have selected this design.

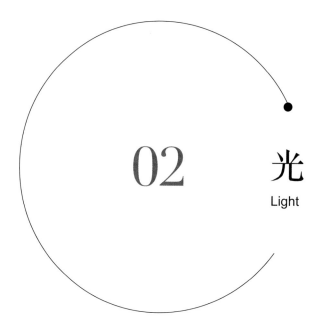

02 光
Light

剪影和影子
Silhouette and Shadow

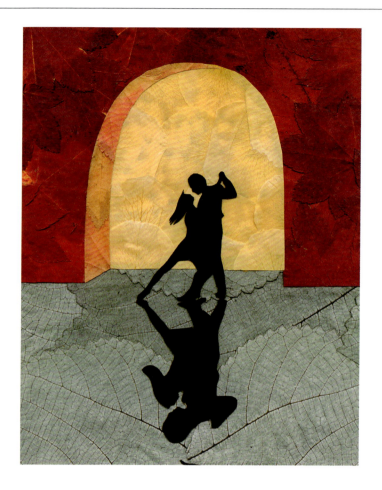

　　剪影最先用于形容一种剪纸，将其贴在背景上，形成鲜明对比的颜色，并经常装裱起来。剪影现在已经扩展到描述一个人、物体或景观，在背光的背景下看起来很暗。任何以这种方式出现的东西，例如，一个站在门口的背影就叫做剪影。

Silhouette images were first used to describe pieces of cut paper, which were then stuck to a backing in a contrasting color, and often framed. Silhouette now has been extended to describe the sight or representation of a person, object, or scene that is backlit and appears dark against a lighter background. Anything that appears this way, for example, a figure standing backlit in a doorway is called a silhouette.

黑色压花材料 Black Pressed Materials

我使用得最多的几种材料是：香蕉皮，孔雀竹芋（calathea lancifolia），花叶青木（aucuba japonica），朱蕉和紫苏。

Several materials that I use the most are: banana peel, rattlesnake plant (calathea lancifolia), aucuba (aucuba japonica), black Ti plant, and shiso (perilla frutescens).

青木 Aucuba

关于如何压制香蕉皮，请参阅我的书《压花艺术（高级）》。一个确保获得非常深色的香蕉皮的方法是冷冻果皮，然后解冻再压。

Refer to Book 3 about pressing banana peels. One sure way to obtain very dark banana peel is to freeze the peel and then defrosting before pressing.

压制孔雀竹芋，最好就是使用微波炉因为普通压花板很难把它压平。

孔雀竹芋 Rattlesnake Plant

For rattlesnake plant, it is best to use a microwave press since it is difficult to get it flat with a regular press.

花叶青木先放在冷冻柜中直到结冰。取出让它解冻和氧化10分钟。用干燥板或微波炉压青木，可以得到很深色的叶子。

Place aucuba leaves in the freezer until frozen. Take them out and let them sit for 10 minutes. Press the aucuba leaves with any method to obtain very dark leaves.

黑色朱蕉和紫苏都可以用干燥板来压。

Press black Ti leaves or shiso with a desiccant press.

剪影设计 Silhouette Design

我们喜欢跳阿根廷探戈。我就想制作一幅跳探戈的画。

Argentina tango is something we enjoy. I want to make a picture about this subject.

首先画一个轮廓图,如果画图有困难,可以打印照片,然后描出轮廓。

First, I made an outline sketch. If you have problems with making a sketch, you can print the picture out and then trace the outline.

影子 Shadow

我们需要先确定光源来决定阴影。阴影不是简单的把剪影反过来而已。

We need to determine the light source in order to draw the shadow. Shadows are not as simple as a reverse of the silhouette.

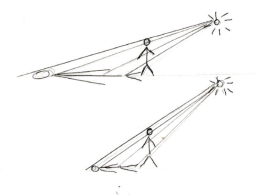

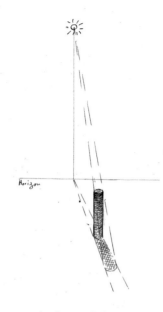

左侧图解绘制阴影的方法。

首先，我们需要确定光源和地平线。然后从光源绘制垂直线到水平线。画出来自地平线上光源下面的点到物体底部的线条。然后从光源画线条到物体的顶部。从物体底部到这些线条相交的点决定阴影的区域。

The way to draw shadow is illustrated on the left.

First, we need to determine the light source and the horizon line. Then draw a perpendicular line from the light source to the horizon line. Draw lines that come from the point on the horizon where it is directly below the light source to the base of the object. Then draw lines from the light source to the top of the object. The area from the base of the object to the points where those lines meets determines the shadow.

背景 Background

为强调焦点背景可以简化。将一对跳舞人放在靠近门口的位置。门外屋顶有明亮的灯光。从较暗的舞场看过去，就只有看到他们的剪影和影子。

Backgrounds can be simplified to emphasize the main focal point. I place the dancing couple near a doorway with light just above the door in the outside hall. Looking from inside of the darker dancing hall, all you would see is the silhouette of the couple and their shadows.

制作画 Making the picture

使用黑色细马克笔来描草图的轮廓，就能够从纸张背面看到轮廓。将双面贴放在草图的正面，然后粘合非常暗的花材以填充整个图像，剪出剪影。

对于阴影，该过程与制作剪影相同，但应该使用稍微浅一点的花材。

Use a black fine point marker to trace the outline of the sketch. You should be able to see the outline from the back side of the tracing paper. Place double sided adhesive onto the front of the sketch. Then glue very dark pressed materials to fill the entire image. Cut out the silhouette.

For shadows, the process is the same as making the silhouette, but we should use pressed materials of a lighter color.

制作背景 Making the background

这幅画的背景制作比较简单，不需要详细介绍了。唯一想说明一下的是黄色。黄色很难保持这样的明黄色，多数黄色一年就开始变成比较淡的黄。这幅作品使用了黄色的底纸，配上黄色玫瑰，色彩就比较浓，长时间保持这样的色彩就完全不是问题了。

The background of this picture is relatively simple and does not require detailed instruction. I do want to explain some nuance about the color yellow in pressed flower art, however. It is difficult for flowers to maintain a deep yellow. Most yellows begin to turn paler after a year. This work uses a yellow backing paper with a yellow rose. The color is strong and maintaining such colors for a long time is not a problem at all with this technique.

光 Light

倒影
Reflection

水中的倒影是由光的反射引起的。湖面比较平静时相当于镜面，对光有反射能力。看到倒影的光线是从水上射来的，而物体反射的太阳光的方向是朝水面向下的。在水面上看，是经过反射后才会改变180°的方向摄入人眼。

The reflection in the water is caused by the reflection of light. When a lake surface is relatively calm, it is equivalent to a mirror surface and has the ability to reflect light. The reflected rays of light seen from the water, and the direction of the sunlight reflected by the object is above the water. The angle in which the light hits the water changes the direction nearly 180 degrees after reflecting to the viewer.

倒影定律 Law of Reflection

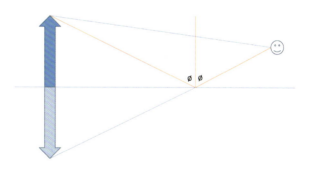

光以相等的角度反射。反射轴垂直于表面。就像这个示例中所画的那样，它有助于勾勒出光线如何撞击物体并与眼睛相遇的示意图。在各条线之间形成相应的角度，这将帮助我们定位对象的反射线。有时，我们必须将其绘制出来，否则几何形状将变得太复杂。但是请记住，光线将始终以相等的角度反射。

Light reflects in equal angles. The axis for reflection is perpendicular to the surface. It helps to sketch out a diagram of how light hits the object and meets the eye, as I did in this example. This gives you angles (Ø) between the various lines that will help you locate the reflection lines of objects. Sometimes we will have to diagram it out. The geometry gets too complex otherwise. However remember that light will always bounce in equal angles.

029

水面的特性 Water Conditions

自然界中几种水的状况，在规划风景画时必须考虑以下几点：

静止的水（没有风时在池塘和小湖中常见）倒影清晰。

水流缓慢（较大的湖泊或流速缓慢的河流）倒影会有一点受水流的影响，不过还算清晰。

水流比较急（在河流和溪流中很常见）倒影只有碎片可见。

浪比较大，看不到反射（倒影）（大风的水域，例如大风天的湖泊和海景）

You can find several conditions of water in nature, which are important to consider when planning a landscape picture:

Still water (common in ponds and small lakes when no wind is present) bodies have vivid reflections.

Water moving lazily (larger lakes or slow moving rivers) create reflections that are somewhat disturbed but still visible.

Water ripples with more motion (common in river and streams) obscure reflections but small pieces of reflection is still visible.

Water so disturbed we can't see reflections (large bodies of water such as lakes and seascapes on a windy day)

反射表面的质地决定了反射的亮度和质量。水总是有一定程度的折射，这意味着倒影会有些偏移。水的涟漪也会使反射弯曲，甚至轻微的波浪也会扭曲并转动图像，从而导致天空和阴影的射入。反射角的变化很大。

The texture of the reflective surface determines the brightness and quality of the reflection. Water always has some amount of refraction, which means the image is skewed a bit. The ripples of the water also bend the reflection. Even slight waves twist and turn the image, reflecting light and shadows randomly. The angles of reflection are all over the place.

这幅画描绘了一个有雾早晨的宁静水面。小船划破平静的湖水。

The picture depicts a still water surface in a misty morning. The small boat rowing across the lake brakes the calmness of the water.

作品材料 Materials

1. 水彩纸（20.5cm×25.5cm）
2. 粉彩（绿叶和天蓝色）
3. 乌蕨
4. 蓍草
5. 红色和橙色的秋叶
6. 紫苏
7. 棕色叶子
8. 白枫杨或苎麻
9. 白芒草
10. 细叶铁线蕨
11. 细小深色根
12. 典具纸
13. 原生态棉花球

1. 8"×10" Watercolor paper
2. Soft pastel (leaf green & sky blue)
3. Squirrel foot fern
4. Yarrow leaves
5. Red and orange fall leaves
6. Dark purple perilla
7. Brown leaf
8. Silver poplar or ramie
9. pampas grass
10. Micro maidenhair fern
11. Small dark roots
12. Tengucho paper
13. Cotton (natural cotton ball)

实物和倒影压花材料 Real and Reflection Pressed Materials

　　湖边的植被是很浓密的。因为是在对岸，所以选择细小叶子的材料。这里选择的是乌蕨。值得注意的是不能全部都是同样绿色的材料，一些枯叶是风景画能够做得自然的必要材料。

The vegetation by the lake is very dense. Select materials with small fine leaves since it is on the opposite bank. I chose squirrel foot fern. It is important to note that not all materials are equally green. Some dead leaves are necessary materials for landscape pictures to look natural.

　　水面倒影的颜色总是比真实物体暗淡一些。因此，我选择了灰色的蓍草叶子。

The reflection is always a little duller in color than the real object in water. Therefore, I have selected a grey colored yarrow leaves.

背景 Background

　　我们需要将背景上色作为画的底色。这一点很重要，以防万一有些区域没有被压花材料完全覆盖而显得突兀。

We will need to color the background as the base color for the picture. This is important because in the case a little bit of area is not covered by pressed materials, it would not look incomplete.

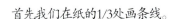

首先我们在纸的1/3处画条线。

First, we draw a line at the top third of the backing paper or canvas.

先使用软粉彩棒的宽面进行着色。用手指揉开并混合颜色。

用绿叶色作为上层1/3。这是茂密的森林生长区。底部的2/3是倒影区域。倒影颜色始终不如真实物体生动鲜艳。我混合了绿叶色、橄榄色和灰色。此外，还展示了小部分的天空。

Use the broad side of a soft pastel stick to color. Use fingers to spread the color.

I used leaf green for the upper 1/3. This is the lush forest growth area. The bottom 2/3 is the area for the lake. The colors of reflections are not as vivid and colorful as the real objects. I mixed leaf green, olive green, and grey colors. Also, I have some small portion of the sky showing.

光 Light

完成着色后喷定稿胶。

Spray fixative when coloring is done.

制作森林和倒影 Making the Forest and Reflection

用乌蕨排列最上面。因为我们描绘森林的下半部分，看不见树顶，所以不要留白。

Use small pieces of squirrel foot fern to line the top. Do not leave empty spaces since we are depicting lower portion of the forest. We do not see the tree tops.

乌蕨从顶部重叠到1/3线。中间夹杂几片深褐色和黄色的叶子，但不要过量。

Overlapping squirrel fern from the top to 1/3 line. Have a few pieces of dark brown and yellow leaves in between but do not overdo it.

倒影则从下往1/3线铺排叶子。把天空露出来，让小山的倒影展现出来。

Work from the bottom to the 1/3 line for the reflections. Expose the sky so the reflection of the small hills is shown.

制作雾 Making the Mist

用一张典具纸覆盖整个画面，以产生朦胧的雾的效果。将棉花去籽撕成薄片，然后排列在下部2/3上。靠近地面的地方应较厚（1/3线），往下逐渐变薄。

Cover the entire picture with a piece of tengucho paper for the misty effect. Remove the cotton seed and tear the cotton to thin pieces and arrange on the lower 2/3. It should be thicker close to the waterline (1/3 line) and gradually thin down toward the bottom.

制作船和人 Making the Boat and People

把图像描两份。在其中一份上，使用马克笔描线，以便可以从背面清楚地看到这些线。将此份用于真实图像。使用另一份做倒影。

将双面贴粘在真实图像的正面。再将双面贴粘在倒影的背面。根据颜色贴叶子并按照图样剪掉多余的部分。

Trace the image twice. On one of the copies, use marker to trace lines so we can see the lines clearly from the other side of the tracing paper. Use this piece for the actual boat. Use the other piece for the reflection.

Place double sided adhesive on the tracing side for the positive image. Place double sided adhesive on the back of the tracing side for the reflection. Glue leaves according to the colors and cut according to the trace.

注意倒影的颜色会比真实的物体颜色浅。

加几小片白芒草为船桨激起的水花。

Remember that the colors of the reflection are lighter than that of the object.

Add small pieces of pampas grass as the boatman's oars strike water.

制作前景植物 Making the Foreground Foliage

使用深色叶子剪下2~3条弯曲的树枝，并将其放在图片的左上角。添加一些小的暗色植物根。然后加入几片迷你铁线蕨作为叶子。

Cut 2~3 pieces of curvy branches and place them on the left corner of the picture. Add a few small dark roots. Then add a few pieces of mini maidenhair fern as leaves.

03 负空间
——落日与水

Negative Space- Sunset and Water

在02章第1节中我详细地介绍了剪影。现在我们来看看负空间，之前使用了带有门框的墙。在本节中，我们将学习落日和水。

有很多方法制作水元素。这是其中一种制作平静的大片海面的方法。

We learned about silhouette in section 2-1. Now we can use the of the negative space technique where we simply just used a wall with door frame in the last section. In this section we will learn about sunset and water.

There are many ways of creating water. This is one way that you can create a large body of calm waves.

我儿子在大学时喜欢冲浪。南加州的海岸不像夏威夷欧胡岛北部那样有很高的海浪，但我们仍然有很多不错的海浪。《归家》是一幅反映加利福尼亚沿海落日的作品。

My son loved to surf when he was in college. Southern California coast does not have very high surf like north of Oahu but we still have pretty good surf a lot of times. "Coming Home" is a picture that reflects California coastal sunset.

材料
Materials

几种不同色彩的秋叶从深红到黄。

黑色材料（参看上一章）。我比较喜欢香蕉皮和青木，因为它们比较有弹性。

Several shades of fall leaves ranging from deep red to yellow.

Black materials (refer to the last chapter). I prefer banana peel and aucuba since they are more flexible to work with.

我们也需要一些同一片叶上有黄色和红色的叶子来获取自然的色彩过渡。

We also need some leaves that contains yellow and red on the same leaf naturally for the color transitions.

  我们居住在南方的人很难找到好的秋天树叶。可以考虑压玫瑰花瓣来制作这种类型的作品。确保也有双色调玫瑰花瓣以达到自然色彩过渡。

It is difficult for those of us that live in Southern California to find good fall leaves. Consider pressing flower petals such as roses to make this type of picture. Make sure to have two tone rose petals as well for natural color transitions.

  最后我们需要一些白色的花瓣。找一些不透明的，我用了白色的非洲菊。

Last we will need some white petals. Find something that will not become transparent. I have used white gerbera daisy petals.

负空间——落日与水 Negative Space- Sunset and Water

绘制图样
Sketch the Picture

　　首先，我们在画面顶部1/3处绘制一条水平线。太阳应该稍微偏离中心。如果愿意的话，画一些远山。接下来我们画出地面和棕榈树的轮廓。最后再画几只飞翔的海鸥。

First, we draw a horizontal line at top third of the page. The sun should be slightly off center. Then draw a distance land mass if you want to. Next we define the ground. Draw outlines of the palm trees. At last we have a few seagulls flying around.

制作作品
Making the Picture

1

把叶子剪成条状。先从地平线的太阳处开始，使用浅黄色贴出半个椭圆形，然后用浅橙色，最后使用深橙色。

Cut the leaves into horizontal strips. Starting from where the sun is on the horizon line, we use light yellow to glue to a half oval shape. Then we have light orange next to the yellow. And then followed by deep orange.

2

继续下半部。不需要覆盖远山和陆地。靠近前面的陆地可以使用比较深色的红。

Continue to the lower portion of the picture. You do not need to cover the distance land and ground. Use deeper color when it is closer to the foreground.

3

取一些白色花瓣（这里使用的是白色非洲菊），剪出一个很小的半圆和一些细小两头尖的花瓣来。

Take some white petals (I used white gerbera daisy there) and cut one small half circle and many tiny pieces with two pointy ends.

将半圆作为太阳粘贴，然后将小片花瓣作为波浪上对阳光的反射贴好。

Glue the half circle as the sun and then the tiny pieces as the reflection on the waves.

负空间——落日与水　Negative Space- Sunset and Water

用橙色、红色和黑色不同颜色剪出更多小片。确保两端尖锐。中间的高度决定海浪的大小。

4

Cut more small pieces with different colors ranging from orange, red, to black. Make sure the two ends are pointy. The more height in the middle indicates bigger waves.

5 从地平线开始，在照片中间粘上比较浅的颜色，当我们往两边离开太阳时，将光度降到到深橙色和红色，最后变为黑色。因为地平线远离海边观察点，所以使用最小的小片。把小片贴得密集，因为从远处望过去，海浪很密。在远山的底下贴几片浅色的，会有一些色彩对比。但是不能很多，因为这边基本上是阴的。

Starting from the horizon, glue lighter color in the middle of the picture, as we move to the sides away from the sun, increase shade to deep orange and red, and finally to black. Use the smallest pieces since the horizon is far away from the view point. Glue the pieces closely together since waves seem very tightly packed far away. Glue a few lighter color pieces at the base of the distance land for some contrast.

继续往下面贴。当我们越往前的时候，使用的"小片"也慢慢加大尺寸。

6

Continue gluing the pieces. As we move closer to the viewer, we use bigger pieces.

7 把"海"用小片贴满。剪两片香蕉皮作为树干。

Fill up the small pieces all the way covering the "sea". Cut two pieces of banana peel to form the tree trucks.

041

剪细条来做树叶的梗。

Cut thin lines to form the leaf branches.

9

剪一些细长的楔子做棕榈树的叶。粘贴棕榈树叶。

Cut some skinny long wedges for the palm leaves. Glue the palm leaves.

用紫苏顶尖上的小叶子做海鸥。

Use the tip small leaves of purple perilla as seagulls.

负空间——落日与水 Negative Space- Sunset and Water

11

用黑色的叶子或香蕉皮铺满海滩。

Use black leaves or banana peels to fill the beach.

12

如果需要,我们可以让冲浪者在沙滩上走回。描这个图像,用它来剪出一块黑色花材并贴在棕榈树下。

If desired, we can place a surfer walking on the beach. Trace the image, use it to cut out a piece of black pressed material and glue under the palm tree.

04 色彩和谐
Color Harmony

色彩和谐定义为"请形成统一整体的色彩排列"。这是色轮。

Harmony in color is defined as a "pleaseing arrangement of colors forming a consistent whole". Here is the color wheel.

有几种色彩组合效果很好。

There are a few color combinations that works well.

互补色
Complementary colors

色轮相反侧的颜色。

colors on the opposite sides on the color wheel.

这样强烈的对比通常用于突出关注中心。

The strong contrast is often used to draw attention to the area of interest.

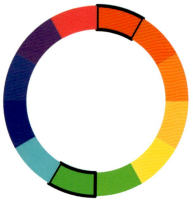

分割互补色
Split Complementary

　　此变体使用一种基色和与色轮上直接相对颜色相邻的两种颜色（如下图标示）。分开的互补色和谐仍然像前面互补色方案一样提供对比度，但是它通过使用两种类似颜色来产生比较柔和的配色。这种和谐的冲突性较小，用途更广。

Instead of using two complementary colors, this variation uses one base color and the two colors adjacent to the directly opposing color on the color wheel. A split complementary color harmony still offers contrast like a standard complementary color scheme, but it creates less tension by using colors analogous to the true contrasting hue. This harmony is less aggressive and more versatile.

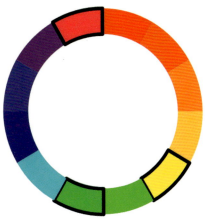

类似色
Analogous

　　类似的颜色在色轮上彼此相邻。这种低对比度的组合是平静的，并且在自然界中经常出现，使其成为天然最和谐的色彩搭配之一。通过选择一种基色并使用两种或三种类似的颜色作为强调阴影

色彩和谐 Color Harmony

来在设计中达到平衡,且尝试仅使用暖色或冷色以避免复杂的和谐感。

Analogous colors are located next to each other on the color wheel. This low-contrast combination is calm and often found in nature, making it one of the most instinctively harmonious color pairings. Create balance in your design by choosing one base color and using two or three analogous colors as accent shades. Try to stick to using only warm or only cool colors to avoid complicating the harmony.

三向轴
Triadic

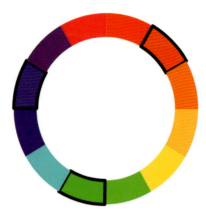

　　三向轴采用在色轮周围均匀分布的三种颜色。三向轴通常是充满活力的，但是不会像互补色方案有时刺目。为了使您的设计保持平衡，让一种颜色占主导地位，而将另两种颜色作为增色。

A triadic color harmony employs three colors evenly spaced around the color wheel. Triadic harmonies are often vibrant, but without the sometimes-jarring look of a complementary color scheme. To help keep your design balanced, let one color dominate and use the others as accent colors.

色彩和谐 Color Harmony

四角色
Tetradic

　　四角色有两种类型：长方形和正方形。这种配色使用两组互补对排列的四种颜色。四角色色彩种类最多，非常引人注目，因此能达到活力四射的效果。但是，由于它们涉及四种不同的颜色，因此需要达到有效的平衡较难。运用四角色的最佳方法是让一种颜色占主导，让其余三种颜色成为次要元素。还应该努力平衡暖色和冷色，以免显得过于花哨。

There are two types of tetradic harmonies: rectangular and square. These harmonies use four colors arranged in two complementary pairs. Tetradic harmonies offer the most variety and tend to be very eye-catching, and as a result work well with flashier subjects. However, because they involve four different colors, it can be difficult to strike an effective balance. The best way to employ a tetradic harmony is to let one color dominate and reserve the three remaining colors for accents or secondary elements. We should also strive to balance warm and cool colors to avoid appearing garish or overwhelming.

色彩和谐 Color Harmony

自由配色
Your Own Color Pallet

以上这些只是供您考虑的一些颜色搭配示范例。大自然是我们最好的老师。我们可以从自然界中汲取灵感，制作属于自己的色彩搭配。在作品《大自然的交响曲》中，我使用了一系列彩虹色的花，创作了丹顶鹤飞过的花田。

All about are just some examples of color pallets for you to consider. Nature is our best teacher. One can come up with own color pallets with the inspirations from the nature.

In "Symphony of The Nature" I used rainbow colors of flowers forming an imagination flower field where cranes fly by.

可以有无限的色彩和谐组合。色轮模型只是我们了解各颜色之间关系的基本指南。

There are infinite color harmony combinations. Color wheel model is just a basic guideline for us to understand relationships between colors.

05 透视
Perspective

我在我的《压花艺术（高级）》书中给出了一个一点透视设计的例子。我收到了读者和学生的许多问题。许多人要求我提供一个有树木的街道为例。在这里，我设计了一条安静的街道，两旁都是蓝花楹树。两侧还栽种着灌木丛。一头小鹿从左边的蓝花楹中探出。在这幅作品中，我还添加了光点。

　　我们每天都看到街道。许多街道绿树成阴。在南加利福尼亚州，每年晚春我们都会看到蓝花楹开花。如果你见过蓝花楹盛开，就会知道它一开花整条街都被紫色覆盖了。

I have given an example of one-point perspective design in my advanced pressed flower art book. I have received many questions from readers and my students. Many have asked me to give an example of a street with trees. Here, I have designed a quiet street lined with jacaranda trees. The two sides are further lined with bushes. A young deer peeks out from the jacaranda trees on the left. In this picture, I also added light spots.

We see streets every day. Many streets lined with trees. In Southern California, we see jacaranda trees bloom late spring. If one ever sees the jacaranda tree bloom, one would understand that the entire street is shaded in purple.

材料 Materials

1. 水彩纸28cm×35.5cm
2. 蓝色，紫色，边缘浅蓝中间白色绣球花
3. 有棕色斑点大蕨叶（或任何橄榄色的带有斑点的叶子）
4. 银白杨树叶
5. 花叶青木
6. 浅褐色的叶子
7. 中褐色的叶子
8. 染紫色和蓝色的蕾丝花
9. 白色蕾丝花
10. 绿色夕雾花

1. Watercolor paper 11"×14"
2. Blue, purple, light blue with white center hydrangea
3. Big fern leaves with brown spots (or any olive colored leaves with spots)
4. Silver poplar leaves
5. Aucuba leaves
6. Light brown leaves
7. Medium brown leaves
8. Dyed purple and blue Queen Anne's lace
9. White natural Queen Anne's lace
10. Green trachelium

一点透视构图 One-point Perspective Drawing

在底纸的下端1/3处绘制一条水平线。定义消失点。这一点不应在正中心。它应该稍微在左侧或右侧。我把它定在右边。画出街道、人行道、矮树丛和蓝花楹树顶等线条。

Draw a horizon line on the lower third of the paper. Define the vanish point. This point should not be on the dead center. It should be either slightly on the left or on the right side. I made it on the right side. Draw the lines for the center road, side walk, bush line, and tree tops.

矮树丛的底色 Color Base for Bushes

矮树丛的底色绿色为上端，下端深色。矮树丛的阴影为深色。

The color base for the bushes are dark on the bottom and green on the top. The dark color is due to the shade of the bushes.

制作街道 Making the street

使用银白杨树叶制作街道。

Use silver poplar to make the street.

制作人行道 Making the Side Walk

我把人行道用有褐色斑的大蕨叶做出郊区土路长一点草的样子。

I made the sidewalk using large fern leaves with brown spots to illustrate the ground covered with dirt and a little bit of grass.

制作矮树丛 Making the bushes

用绿色夕雾花从中间往外，从上往下，层叠贴出矮树丛的模样。

Use the green trachelium from the middle toward the sides, from the top toward the bottom, to layer the appearance of bushes.

制作树顶底色 Making the Base Color of the Tree Tops

使用紫色和蓝色绣球花填充天空。在街道天际使用浅色。使用各种大小的绣球花，不要充满整个天空，这里只是制作底色。

Use purple and blue hydrangeas to fill the sky. Use light color on top of the street. Use a variety of sized hydrangea. Do not fill the entire sky. This is just for the color base.

制作树干和树枝 Making the Tree Trunk and Branches

将树干排成行。请记住,远处的树木看起来树木之间距离小,看起来矮小。当它们离我们越来越近时,它们变得越来越大,树干更粗,并且树木之间距离也加大。每个树干也应有所不同,而且它们都不是笔直的。请勿使用一个模板做所有的树干。

Line the tree trunks. Keep in mind that the trees would seem closer and smaller as they are far away. They become larger, thicker, and spread apart as they are closer to us. Also, each tree trunk should be a little different. They are never completely vertical. Do not use a template for all of the tree trunks.

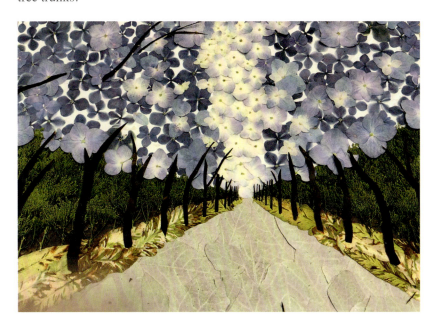

人们将能够看到较近树干的明暗。我们可以将中等和浅褐色的细条粘到深色树干上,使它们看起来更具立体感。

One would be able to see the light and shade of the tree trunks closer to the viewer. We can glue thin strips of medium and light brown leaves onto the dark tree trunks so they look more dimensional.

注意:我这幅画的光源来自左边。因此浅色的条要贴在所有树干的左边。

Note: I have defined that the light is coming from the left side. Hence all of the light color strips are to be glued on the left side of the tree trunks.

在上面增加树枝。

Add more branches on the top.

制作蓝花楹的色彩变化和光点 Making the Color Variation and Light Spots of Jacaranda

因为光源来自左边，因此照在蓝花楹上的光点就在右边。细调绣球，紫色和蓝色蕾丝花让人看起来整条街的上空都被蓝花楹覆盖。要让天空留有一些小空隙。

Since light comes from the left, the light spots on the flowers are cast on the right side. Adjust the hydrangea base color, purple and blue Queen Anne's laces so one would see the entire sky of street almost covered with jacaranda flowers. Let the sky having some small uncovered spots.

在街道上撒一些紫色蕾丝落花。

Add some purple Queen Anne's laces on the street for falling flowers.

加入小鹿增加画作的有趣度 Adding The Young Deer for Visual Interest

用花叶青木刻一个小鹿。把它安排在消失点附近。用描图纸把小鹿的形状描出来。把双面贴贴在描图纸的背面。贴上黑色叶子，然后把小鹿剪出来。

Use aucuba leaf to carve a small deer. Add it close to the vanish point. Use double sided adhesive on the back of traced image. Glue a dark colored leaf on the top and cut out the deer.

06 风景画设计

Landscape Design

关于风景设计：运用自己的想象力。除非我们制作一幅名胜景观，不需要像照片那样把画作成和真实的场景一模一样。可以将真实场景的元素组合在一起并重新排列以制作出自己的艺术。

About landscape design: use our imagination. We do not need to make it as real as a photograph unless we are making a famous landmark. It is absolutely alright to combine elements from real places and rearrange them to make our own art.

主题
Theme

首先要确定设计主题。要表达什么？这幅画我的主要目的是展示一个小村庄和这个村庄的自然环境。最初的景观研究是有石墙的苏格兰。我又添加了同样有石墙的中国水乡元素。东西方的融合使得这幅画非常独特，并吸引了许多人。

First we need to determine the theme of the design. What do you want to express? In this picture, my main goal is to showcase a small village and the natural setting of this village. The original landscape studies were Scotland with stone walls. I have added elements of villages found in South East China, which also have stone walls. The blend of East and West make this picture very unique and is a draw to many.

三分法则
Rule of Third

正如我们之前所知，为了使风景画具有深度，我们将画的3个纵切立体部分（背景、中景和前景）呈现在一个平面中。这被称为三分法则。背景远离视点，前景最接近视点。

As we have learned before that in order for a landscape picture to have depth, we present the picture in 3 planes – background, middle ground, and foreground. This is referred as the rule of third. Background is far away from the view point and foreground is closest to the view point.

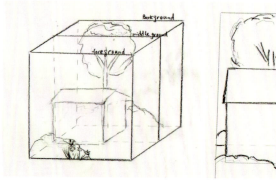

这个简图描绘了我们如何将立体的物件分割出3个不同的板块呈现于从前面看过去的平面图上。

This illustrates how we slice a three dimensional landscape and present it in a two dimensional picture that we view in the front.

确定焦点
Decide Focal Point

有趣的是，我们的焦点并不总是在前景，有时它可能处于中景。

It is interesting that our focal point is not always on the foreground. Sometimes it can be on the middle ground.

构图
Composition

构图是一种组合设计元素以达到美的目标的艺术。没有将自己的感受注入的情况下仅仅在制作照片似的画面是不够的。艺术就是将我们的感情传达给观众。

Composition is an art of combining design elements to reach aesthetic goal. It is not enough to make a photo realism picture without injecting your own feelings into the work. Art is about communicating our feelings to the viewers.

我经常勾勒出一些缩略图来决定我的构图。绘制这些小图很容易让你看看是否喜欢这些形状。一旦确定了基本构图，就可

以将构图绘制到合乎比例的图上。

I often sketch out some thumbnails to decide on my composition. It is easy to sketch these small images and see if you like the shapes. Once you decide on the basic composition, then you can draw the composition to the scale.

没有必要绘制与画作完全相同的尺寸。只要将图片绘制到合乎比例，稍后在电脑上将其放大。对于较大尺寸的作品尤其如此。

It is not necessary to sketch the same exactly size as the final picture. As long as we draw the picture to the scale, we can enlarge it on your computer later. This is especially true for larger size works.

透视和尺寸比例必须是合理的。参看我在《压花艺术（高级）》中介绍的一点和两点透视。

It is important that the perspective and dimensions are reasonable. Review one-point and two-point perspective that I have covered in Book 3.

画出越多细节，最后的压花画就会越好。画图能让我们在脑中思考想表达什么，比例、光影等等。在这一步，还可以思考要使用什么花材。

The more details you put into our drawing, the better final result will show. Drawing is the step that we plan in our mind about what we want to present, proportion, light and shadow, and etc. In this step, we actually think about the botanical materials you will use.

在这个作品中，我使用了两点透视来展现不同的建筑元素以及它们之间的关联。

In this composition, I have used two points perspective to showcase the different architectural elements and the relationships between them.

建筑元素技巧
Archtectural Element Techniques

材料 Materials

1. 薄绢或纸（20.32cm×25.4cm）
2. 白卡纸（20.32cm×25.4cm）
3. 双面贴
4. 白菜叶或任何青柠色的叶子
5. 欧芹或大铁线蕨
6. 黑叶朱蕉或香蕉皮
7. 苎麻或醉鱼草叶或银覆盆子叶
8. 白桦树皮或白千层树皮
9. 黄色香雪球或任何小黄花
10. 翠云草或任何苔藓
11. 小棕榈叶或任何长而窄的叶子
12. 婴儿泪或任何小叶子
13. 紫雪茄或任何小的粉紫色花朵
14. 白色飞燕草或白色绣球花
15. 紫色夕雾或紫色香雪球
16. 小町藤或任何小紫色花
17. 嫩紫藤叶、藤蔓或其他藤蔓，如千叶吊兰
18. 迷你铁线蕨或任何小叶子
19. 笔仔草（金丝草）、白芒草或其他细长草籽
20. 紫苏
21. 茶包或小片典具纸

1. Thin silk or tengucho paper (8"×10")
2. White card stock (8"×10")
3. Double sided adhesive
4. Napa cabbage leaves or any lime colored leaves
5. Parsley or maidenhair fern
6. Black Ti leaf or banana peel
7. Ningma or butterfly bush leaves or silver raspberry leaves
8. Birch bark or paper bark tree bark
9. Yellow alyssum or any small yellow flower
10. Peacock moss (Selaginella uncinata) or any moss
11. Small palm leaves or any long and narrow leaves
12. Baby tear or any small leaves
13. Mexican heather or any small pink purple flowers
14. White larkspur or white hydrangea
15. Purple Trachelium or purple alyssum
16. Hardenburgia or any small flowers
17. Young wisteria leaves and vine or other vines such as wire vine
18. Mini maidenhair fern or any small leaves
19. Pogonatherum crinitum or pampas grass
20. Dark purple perilla
21. Tea bag or small piece of tengucho paper

制作步骤 Procedure

我们把繁琐的画面解开成单个的元素,先把每个元素一一制作出来。

We break down a complicated picture into elements and work on one element at a time.

我们可以通过电脑扫描并且将图调整到和画面一样大小来减轻制作的负担。我们还可以翻转图像,让我们有两个反向图。便于制作时修剪形状。

One thing that can make our lives easier is to scan our drawing and size it to the same size as our planned picture with a computer. We also flip the image horizontally so we have two reverse images that we can use to trim shapes with.

房屋 Houses

1

我们把一张反转图像的房屋墙体剪下来。每面墙都剪开。

We cut out the house walls from one of the reverse images. Separate each wall.

2 在朝阳的墙壁上使用白色的飞燕草花瓣。使用较小的花瓣（如飞燕草或绣球）会比使用较大的平整花瓣（如玫瑰）更好地模拟墙壁的灰泥质地。修剪朝外的一面。不必修剪另几侧，因为它们会被其他材料覆盖。

Use white larkspur petals on the wall that light shines on. Using smaller petals such as larkspur or hydrangea would simulate the stucco texture of the wall better than a large piece of smooth petal such as rose. Trim the side that face out. It is not necessary to trim the other sides since they will be covered by other materials.

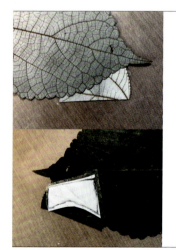

3

使用两种不同程度的灰色苎麻叶来制作一些光影变化让画面更有趣。翻过来把形状剪出来。

Use two different shades of ramie leaves to create shade variation making the picture more interesting. Flip over to trim the shape.

4

加一小条深色叶子，让光影对比更强烈。

Add a small strip of dark leaf to make the contrast of light and shadow even stronger.

5

从反转的图案中把房顶剪出来。把黑朱蕉叶顺着房顶的坡度贴在房顶图案背面。房顶两边都要顺着坡度，在屋脊处相交。把房顶按照图案剪出来。

Cut out the roof shape from the reverse image. Glue black ti leaf on the back side of the roof. Remember to glue the leaf with vein following the slope of the roof. The leaf needs to joint on the top. Trim to the shape.

6

把房顶安装好。做一个简单的窗但有一些阴影的变化。把窗贴好。

Place the roof on top of the house. Make a simple cutout window with some light and shadow details and glue it in place.

7

其他的房子都是同样制作方法。

Make all the other buildings using the same method.

石墙 Stone Wall

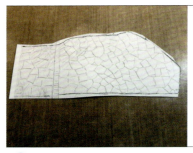

1

把石墙从反转的图上剪下来。

Cut the wall out from the reversed print.

2

把双面贴粘在墙体的白色一面。

Apply double sided adhesive onto the blank side of the wall cutting.

3

把双面贴的离型纸掀开,把墙体贴满黑色的叶子。

Remove the protective paper from the double sided adhesive. Glue black leaves onto the wall.

4

反转过来,把四边多余的黑色叶子剪掉但留一点边。

Flip over and trim off excess material from the sides but do not cut too close to the edge.

5

把苎麻叶剪成不规则小片。

Cut the ramie leaves into pieces.

风景画设计 Landscape Design

6

将苎蔴叶片如同拼马赛克般粘在黑色墙体上。然后翻转根据图纸来修剪墙的形状。

使用镊子将黑色材料的小块弄碎,然后将其粘在石墙块的左边边缘。

Glue the ramie pieces like a mosaic. Then flip over to trim the wall according to the drawing.

Use tweezers to break small pieces of the black material and glue them to the left edge of the stone wall piece.

7

使用镊子尖把边缘的黑色剌成碎片。

Use the tip of the tweezers to chip the black area.

8

打开一个茶袋或拿一小片典具纸。使用湿的小毛笔描小片墙的外部边缘。这是台阶后面的墙,上面有木栅栏。在描痕没干时,撕出形状。

Open a teabag or take a small piece of tengucho paper. Use a wet brush to trace the outside of the small edge of wall. This is the wall behind the stair with the wooden fence on the top. Tear the shape out when the trace is wet.

075

9

在小片墙的接口上涂少量胶水并固定薄纸。将小片墙粘在大墙体后面。

Apply a tiny amount of glue on the side of the small wall piece and secure the thin paper. Glue the small wall piece behind the larger wall piece.

背景 Background

1

首先把绢熨烫平。如果你使用典具纸，打算真空密封，要先把典具纸团几次，再仔细打开。
用浅绿色的大白菜叶按照图纸贴成高光的树顶。
注意，我们是在制作一幅春景。春天树尖上嫩叶是黄绿色的。然后贴上铁线蕨或欧芹叶制作成房屋后面的树。

Iron the silk background piece to remove wrinkles. If you are using tengucho paper instead of silk, and you plan to vacuum seal, then crumple up the paper several times and open it carefully.

Arrange light green napa cabbage leaves to form the highlighted trees on the top of the card stock according to the drawing. Notice that we are illustrating a spring scene. The young tree leaves have a yellow/green shade. Arrange parsley or maidenhair fern leaves to form the trees behind the buildings.

2

把左边的建筑贴好。沿着建筑的地平线贴上矮的绿色叶片和小花。

Place the buildings on the left. Glue some low greens and tiny flowers around the ground of buildings.

风景画设计 Landscape Design

将大白菜叶贴在右边。因为我们想在植物的间隙中看到一点光,所以此处不要使用过深的绿色。

Glue napa cabbage leaves on the right side. Do not use deep green colors here because we want to see a little bit light through the plants.

3

中景制作 Middle Ground

1

用绢或典具纸覆盖背景。将建筑物放在中间。在右侧贴上夕雾或类似材料为建筑物后面的高灌木丛。

无需特别粘贴绢或典具纸。贴建筑物和夕雾足以将其固定住。

Cover the background with silk or tengucho paper. Place the buildings in the middle. Glue some trachelium or similar foliage on the right side to depict tall bushes behind the buildings. There is no need to glue down the silk or tengucho paper. Gluing the buildings and trachelium is enough to hold the material down.

2

把石墙贴好。用一片叶子贴在石墙左边角(绢背后)形成石墙后面的地面。

Glue the stone wall. Glue a small piece of "land" (leaf) on the card stock (behind silk) on the bottom left of the stone wall to indicate there is land mass behind the stone walls.

3

在石墙上增添木栏杆。

Add the wooden fence on the stone wall.

077

近景制作 Foreground

1 在石墙上添加下垂的开花藤蔓。由于这幅画是关于建筑与其自然环境的关系，因此花朵的颜色不要太亮。用翠云草贴"草地"。我们需要掰出翠云草小叶子，一点点地贴，以使其看起来像草。

在绢背面添加一些倒影的提示。我们不需要做逼真的倒影这样会将注意力从建筑物和石墙上转移开。

Add trailing flowering vines on the stone wall. Make sure the flower color is not too bright since this picture is about the relationships of the buildings and their natural environment. Create a grassy riverbank with spike moss. We need to break the small leaves of the spike moss and glue each individually in order to make it look like grass.

Create reflections in the river behind the silk. We don't need to make realistic reflections that distract focus from the buildings and the stone wall.

2

在右手边加一棵树。用迷你铁线蕨做叶子。在树前加上真的带草籽的草。

Add a tree on the right side. Use mini maidenhair fern as leaves. Add real grass with seed heads in front of the tree.

3

最后在左边加一点垂态的开花藤。不要很强烈的色彩，以免和房屋以及石墙争夺注意力。

Lastly, add some trailing flowering vines on the left. Make sure not to overpower the mid ground buildings and stone wall.

基本风景画元素技巧
Basic Landscape Element Techniques

我用这幅压花景观设计的大师级作品来详细讲解一下技巧。我将涵盖以下内容：

蓝天白云

山

悬崖

三种树木

岩石

流水

I will cover some basic landscape elements with this master level project so you can understand the techniques. I will cover the following:

Sky with white clouds

Mountains

Cliffs

Three types of trees

Rocks

Stream

材料 Materials

28cm×35.6cm水彩纸

铅笔

墨水笔

双面贴

描图纸

宠物钢针毛刷

鼠标垫

压制材料：

1. 绣球（白色、蓝色、绿色）
2. 绿色白菜（或羽衣甘蓝）叶子
3. 圆白菜叶
4. 红叶生菜
5. 紫色裂叶羽衣甘蓝（可选）
6. 迷你铁线蕨
7. 黄色的秋叶
8. 白桦树皮
9. 白千层
10. 有斑点的绿叶
11. 绿叶（四种颜色：深色、浅色、中间色、橄榄色）
12. 染深绿色乌蕨
13. 斑点青木
14. 翠云草
15. 棉花
16. 小花（黄色的香雪球，染黄色蕾丝，小蛇目菊，鼠尾草）

11"×14" watercolor paper

Pencil

Pen

Double sided adhesive

Tracing paper

Pet brush

Mouse pad

Pressed materials:

1. Hydrangea (white, blue, blue/green)
2. Green napa cabbage (or kale) leaves
3. Regular cabbage leaves
4. Red leaf lettuce
5. Purple ornamental kale (optional)
6. Misty cloud maidenhair fern (mini size maidenhair fern leaves)
7. Yellow fall leaves
8. Birch bark
9. Paper bark tree bark
10. Spotted green leaves
11. Green leaves (four shades: dark, light, medium, oliv)
12. Dyed dark green squirrel foot fern
13. Aucuba
14. Selaginella (spike moss)
15. Cotton
16. Small flowers (yellow alyssum, dyed yellow queen Anne's lace, small coreopsis, Russian sage)

计划
Planning

用电脑把绘制的图放大到需要的尺寸。如果尺寸大过打印纸，我们可以将图分成两张或更多纸张。

Scale the drawing to size with image editing software. If the size is too large to print, we can separate the drawing into two or more pieces.

风景画设计 Landscape Design

1

我把图片大小调整成可以使用常规打印机上打印的（A4）。根据画面将两张纸粘贴在一起以形成28cm×35.6cm的图纸。使用普通铅笔在图纸的背面进行涂抹（仅绘制图形的轮廓，例如山峰，峭壁和溪流）。

I have scaled the pictures to be printed on a regular printer (letter size or A4). Tape the two sheets together according to the drawing to form a 11"×14" drawing sheet. Use a pencil to scribble the back of the drawing (just the outline of the drawing such as the mountain, the cliff, and the stream).

2

把图放在水彩纸上，用力描画的轮廓，铅笔印子会拓在水彩纸上。

Place the drawing on top of the watercolor paper. Trace the outline of the drawing with some force to create an indent the watercolor paper.

083

天空 Sky

1

把双面贴贴满底纸的天空部分。

Apply double sided adhesive to the area for the sky on the background paper.

2

一点点地揭开离型纸，贴上绣球花瓣。

Lift a small section of double sided adhesive protection paper and glue hydrangea petals at a time.

3

用有白心的绣球花瓣做白云的边缘。白云中间用白色绣球花瓣。

Utilize hydrangea petals with white centers to form the edge of a cloud. Use white hydrangea petals for the middle of the cloud.

4

云应该是自由形态，不要使边缘对齐得太完美而形成一个既定形状。

The cloud should be formless, do not make the edges to line up too perfectly to form a shape.

山 Mountains

1 把圆白菜叶贴在左上的山上，可以用几片叶子来填满整片山。注意一点就是脉都需要合理地朝一个朝向。

Glue cabbage leaves for the mountain on the top left. We can use several pieces to fill the shape. Just make sure the veins run in the same direction.

继续把绿色大白菜叶按照纹理贴中央山脉。用圆白菜叶贴右边山。 **2**

Use green napa cabbage leaves for the mountain in the middle. Notice how the vein's orientation is vertical. Use regular cabbage leaf on the right side.

3

我们用红叶生菜做近处的山。

Next, we will make hills with red leaf lettuce.

用深绿叶子制作右边高地。

Make the highland on the right side using dark green leaves.

5

可选步骤：用紫色裂叶羽衣甘蓝覆盖在高地上。

Optional step: cover the highland with ornamental purple kale.

远处的山色淡，看不见什么细节。距离我们越近，看到的细节越多，色也会比较深。

Distant mountains would seem lighter in color and with little details. As the distance from the mountains gets closer to us, we would see a clearer image with more details and the colors would be darker.

悬崖 Cliff

1 把左边悬崖形状描下来，剪开。上面覆盖双面贴。

Trace the cliff on the left's shape and cut them apart. Cover with double sided adhesive.

2 悬崖的陡峭部分是由黄色的叶子制成的。注意脉的朝向是与地平线垂直的。

The steep part of the cliff is made with yellow leaves. Make sure the veins are vertical as respective to the horizon.

3 这是前面一块悬崖。也用黄色叶子覆盖，按照形状剪去多余的叶子。

This is the front piece of the cliff. Also use yellow leaves to cover and trim to the shape.

4 用绿色叶覆盖悬崖上面的平地。按照形状剪去多余的叶子。

Cover the flat area of the cliff with green leaves. Trim to the shape.

风景画设计 Landscape Design

5

用小片胶带把陡峭的悬崖部分和上面平地连接起来。注意不要有空隙。

Use small pieces of tapes to connect the steep cliff and the flat area together. Make sure there is no gap in between.

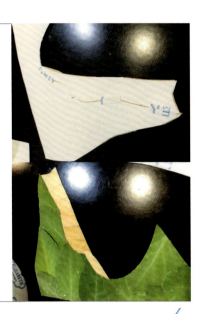

6

把悬崖的脱落部分面朝下贴在白千层树皮的背面。剥白千层，用中间的，正面像沙子的那层。

Glue the drop off area of the cliff face down on the back side of the paper bark tree bark. The front side looks very much like sand.

水和岸背景 Water and Shoreline Background

1

描出图纸上水的形状如右。用双面胶覆盖。然后将蓝色/绿色绣球粘在整片上。把地平线修剪平坦。

Trace the river's shape onto a piece of paper. Cover with double sided adhesive. Then glue blue/green hydrangea on the entire piece. Make the horizon flat by trimming excess material.

2

把悬崖和水面固定在底纸上。水面和岸连接的地方不要贴死。

Assemble and glue the river and cliff. Do not totally glue down the horizon on the river.

3

斑点的叶子把中间的粗脉剪掉分为两半。排列半片叶子形成锯齿形的岸。

Cut spotted leaves in the middle. Get rid of the large main vein. Arrange the leaves to form a zig-zag shoreline.

岩石 Rocks

1

有一些叶子天然可以用来制造岩石。花叶青木在炎热的夏天阳光下暴晒后（收集较旧的叶子），然后放入冰箱冷冻直至结冰。解冻后压。

There are a few leaves that are good for making rocks. We can use aucuba exposed to the hot summer sun (older leaves) and then placed in freezer. Thaw and press.

2

我们可以使用一些受损的叶子来制作微小的岩石。这些叶子的边缘自然粗糙，这会给这块小石头带来光影效果。

For tiny rocks, we can use some damaged leaves as shown. These leaves have naturally rough edges which create realistic looking rocks.

风景画设计 Landscape Design

3 准备三种明暗度不同的棕色叶子，浅色、中间色和深色来制作较大的岩石。用镊子将叶子撕成所需的大小和形状。不要使用剪刀，因为我们需要粗糙的边缘才能使岩石看起来自然。将中间色的叶子压在浅色的叶子上，让浅色于右上方探出。将深色叶子压在左下方，以形成具有光和影的岩石。

For larger rocks, prepare three shades of brown leaves, light, medium and dark. Use tweezers to tear the leaves to the size and shape needed. Do not use scissors because we need the rough edges to look natural. Place the medium shade of leaf over the light one but allow the some of the lighter one to show on the top right edge. Place the dark shade of leaf over on the bottom left to form a rock with light and shadow.

4 在河流两边布置石块。需要制作不同大小和形状的岩石。

Arrange the rocks on two sides of the stream. Make sure you have rocks of different sizes and shapes.

树木 Trees
松树 Pine Tree

1

把花叶青木的主脉剪下来。使用叶尖部分这样脉不会过于粗大。

Cut the main vein of the aucuba leaf out. Use the top portion of the vein so it is not too thick.

089

2

用镊子掰出细长的条，贴在脉上面。

Use tweezers to break some thin strips of the aucuba leaf. Glue the strips onto the vein as shown.

3

继续贴出树的模样。注意，不要形成很完美规则的三角型。

Continue to glue the strips to form the tree's shape. Note: do not form a perfect triangle.

4

用乌蕨来覆盖树枝。

Use squirrel foot fern to cover the branches.

5

我们可以直接使用乌蕨来制作远距离的松树。需要把乌蕨底部插在水面背景的下面。

We can use squirrel foot fern for distant pine trees like shown below. Insert the bottom of the fern onto the river's background.

6

把松树根据底稿来进行安排。

Arrange the pine trees according to the drawing.

迎客柏树 Wind Sculpted Cypress

7

用铅笔在描图纸上描出迎客柏树的枝干。然后翻转过来用墨水笔描。

Trace the wind sculpted cypress truck and branches with pencil. Then flip and mark the reverse side clearly with pen.

把双面贴贴在描图纸的正面（铅笔印的一侧）。揭开离型纸，把香蕉皮贴好。

2

Apply double sided adhesive on the front side of the trace (pencil side). Take off the protective sheet and glue banana skin onto the trace.

3

根据墨水笔描出来的痕迹把树干和树枝剪出来。

Cut the tree truck and branches out according to the pen markings.

用镊子把棕色叶子掰成细条。然后贴在树干的中间偏右侧。

4

Use tweezers to tear thin strips of medium shade brown leaf. Glue the strips onto the middle right of the truck and main branch.

5

用镊子把黄色或浅咖啡色叶子掰成细条。然后贴在树干的右侧。

Use tweezers to tear thin strips of yellow or light brown leaf. Glue the strips onto the right of the truck and main branch.

风景画设计 Landscape Design

6

选择深浅不同的3个绿色叶。用宠物钢针毛刷在鼠标垫背面刺出小洞。

Select three shades of green leaves. Use the pet brush to punch leaves on the back of a mouse pad.

7

用镊子夹出碎片来制作柏树叶。先贴深色的,然后中间色,最后贴浅色。不要把叶子贴得太密,需要有些间隔。柏树的叶子从远处观看是一丛丛,之间有空隙的。

Use tweezers to break small pieces of the leaves to form the leaves of the cypress. Place the dark ones first, then some medium shade, and finally the lighter green ones. Do not glue the leaves too densely. Leave some space since the cypress leaves form groups and appear airy when viewed from afar.

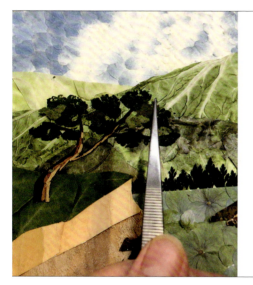

桦树 Birch

1

双面贴一面贴在白纸上(6.5cm×16.5cm)。另一面贴上桦树皮。

Apply double sided adhesive onto a piece of paper (2.5"×6.5"). Glue birch bark onto the paper to form a small sheet of birch bark.

093

2

把树干剪出来。不要剪得笔直。如果徒手剪太困难，可以在背面的白纸上先画出来，然后再剪。

Cut out the tree truck. Do not cut perfectly straight lines. If cutting free handed is too difficult, you can draw on the back of the adhesive side first and then cut to shape.

3

使用镊子把橄榄色叶子掰成细条。

Use tweezers to snap olive color leaves into thin strips.

4

把细条贴在树干的左边。

Glue the strips to the left side of the tree truck.

5

翻转，修剪树干边缘。

Flip and trim the edge of the tree truck.

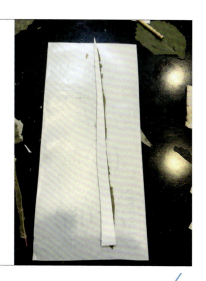

6

把桦树干贴好，并且加上树枝。用迷你铁线蕨做叶子。

Arrange the birch trees into groups. Add a few tree branches. Use mini size maidenhair fern as leaves.

草原高地 Grassy Highland

使用翠云草来覆盖高原。这步可跳过,但是制作这步会对整体画的细节增色不少。

Use spike moss as grass to cover the highland. This step is optional but would add details to the picture.

野花 Wild Flowers

使用蕾丝的单个细小花做为远处的野花。

Use individual Queen Anne's lace floret as wild flower from afar.

风景画设计 Landscape Design

2 中景我们用飞燕草叶子做草,配上黄色香雪球为野花。

We have some tall grass made with larkspur leaves and wild flowers made with yellow alyssum in the mid-ground.

3 近景野花,可以使用稍大一点的花材,但不能所有花材都大,必须有大有小,色彩搭配合适。

We can use slightly larger wildflowers in the foreground. However, we need to select one color theme and mix of large and smaller flowers.

溪流 Stream

因为化妆或医用棉球中有小的白色颗粒,所以最好使用纯天然棉花。将棉花搓扭成条状,如上左图所示。如上右侧图排成一行形成水流。

It is best to use real cotton since there are small white particles in the cotton ball. Twist cotton into a strip as shown. Line the strips to indicate the direction of water flow.

平静的水面 Calm Water

1. 把双面贴一面贴上白纸，另一面覆盖白色花瓣。任何白色花瓣都可以。剪出细小枣核形。

Apply double sided adhesive onto white paper and then glue white flower petals onto the other side. Any type of white petals would work. Cut very small pieces with two pointy sides.

2. 把小枣核形贴在远处水面上形成小波浪。

Glue the small pieces to the large body of water to form some small waves.

3. 完成。

finish.

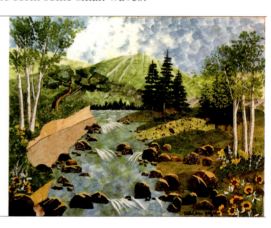

风景画设计 Landscape Design

压花瀑布
Pressed Flower Waterfall

　　本篇将教授几种制作岩石的方法以及如何使用压制的植物材料制作瀑布，我们还将讨论风景设计中的明暗和透视。本篇也教授如何制作立体的枝叶。

This section covers several ways of making rocks and how to use pressed plant materials to make waterfalls. We will also talk about light and shadow, and perspective in landscape design. This class will also teach about how to make three dimensional looking foilage.

材料 Materials

1. 28cm×36cm水彩纸
2. 23cm×30cm黑色卡片纸
3. 黄色、深绿色、橄榄色和黑色粉彩
4. 30cm×40cm 典具纸或丝纸（可免）
5. 1~2朵天然棉花（可免）
6. 棉花球（普通家用）
7. 美人树绒毛、马利筋绒毛或竹纤维（可免）
8. 压制的黑色、深棕色和深绿色青木叶（或这些颜色的任何叶子）
9. 压制的鳄梨壳
10. 天然深绿色和浅绿色（新生长）的铁线蕨（选择较小的叶片类型）
11. 细叶白雪木（或任何小的白色或黄色花）
12. 压的叶子（浅、中、深棕色）
13. 小片透明亚克力（可免）

1. 11"×14" watercolor paper
2. 9"×12" black cardstock
3. Soft pastel in yellow, hunter green, olive, and black
4. 12"×16" tengucho or sanwa paper (optional)
5. Natural cotton ball (optional)
6. Household cotton ball
7. Silk floss tree floss or milkweed floss or bamboo fiber (optional)
8. Pressed aucuba (or any leaves) in black, dark brown, and dark green
9. Pressed avocado shell
10. Maidenhair fern (select the smaller leaves) in natural deep green and light green (new growth)
11. Euphorbia diamond frost (or any small white or yellow flowers)
12. Pressed leaves in light, medium, and dark brown
13. Small piece of clear acrylic (optional)

压制鳄梨壳窍门 Tips on Pressing Avocado Shell

1

将果肉从壳中刮得很干净。将一个壳垂直切成四份。

Remove the edible portion from the shell. Cut one shell into four pieces, vertically.

风景画设计 Landscape Design

然后在每个四分之一壳的两边外侧切一些小口，这样可以将壳压平。

Then cut some small slits on both sides of each quarter so the shell can be pressed flat.

3

将纸巾垫在壳下，用微波炉压。

Place a paper napkin under the shell and press with a microwave press.

4

高温下压制三次，分别为20秒、20秒、40秒。按照壳的状况调整最后的时间，毕竟每个微波炉不同，壳的情况也不同。最初不要压超过20秒以避免过热引起的水泡。每次都让蒸汽散出。将外壳放入压花板或旧厚书中以完成干燥。

On high power, microwave three times for 20, 20, and 40 seconds for a large avocado shell. Adjust the last set of time according to the size and condition of the avocado shell since not all microwave ovens are the same. Do not press for longer than 20 seconds initially to avoid blisters from overheating. Let the steam out between each time. Place the shell in a press or in an old phonebook to finish drying.

压制花叶青木窍门 Tips on Pressing Aucuba

1 获取全黑叶片：把叶片放在冷冻库中冰冻至硬。取出解冻摆放10分钟。然后可以用任何方法压制。

Obtaining black leaves: place the leaves in the freezer and freeze them completely. Take them out to defrost and wait for 10 minutes before pressing. They can pressed in any type of press.

2 获取橄榄的黑色叶片：从冷冻柜取出的叶子立即用微波压花板压。叶片会压干并氧化至由橄榄色至黑色。

这些叶子是微波压制的。从左边开始，依次为从冷冻库取出10分钟后压、从冷冻库取出立刻压制、没有冷冻过的。

Press and microwave frozen leaves with a microwave press to obtain olive to black colored leaves. The leaves will oxidize into black to olive colors.

These leaves are microwave pressed. From left to right: 10 minutes of thawing, immediately after taken out of the freezer, not frozen.

创作理念 Concept

几年前我前往伊瓜苏瀑布，在阿根廷和巴西两边的所有小径上走来走去，观看和研究瀑布——那么多大大小小的瀑布。这幅画是关于我心目中的热带瀑布。

I traveled to Iguazu Falls and walked up and down the trails in both the Argentine side and the Brazilian side to view and study waterfalls. There are so many! Large and small. This picture is about those tropical waterfalls in my mind.

风景画设计 Landscape Design

关于光线和环境 About Light and Environment

这幅画的光源来自上方。这是一个窄细峡谷之中的双瀑布，巨大的石头形成峡谷壁，瀑布下方的湍流奔腾向左下游流去。光从上面透过树梢撒落一点到峡谷里，石壁潮湿，一些蕨类和青苔在石缝中生长。瀑布中的水汽在山谷中形成雾状。

The light source for this picture comes from the top. There are twin waterfalls in a narrow canyon. Huge stones form the cliff walls. The turbulent water below the falls rush down to the left. Light is sprinkled from the top of the tree into the river below. The stone walls are wet. Some fern and moss grow between the cracks. The water vapor from the falls form a misty atmosphere in the canyon.

制作背景 Making the background

将软粉彩侧面涂在水彩纸上。黄色应该在纸的左2/4左右。

Place the soft pastel sideways and color the watercolor paper as shown. The yellow should be offset slightly to the left, rather than in the middle (2nd quadrant).

2

用手指把颜色涂开。如果需要的话，可以在上面增加颜料。换色的时候记得洗手。

Use your fingers to spread the colors. Add more color when needed. Wash hands when switching colors.

3

颜色边缘的地方注意需要自然过渡如图。

Blend colors as shown.

4

完成的背景颜色。
喷定稿剂固色。

Finished background coloring.
Spray fixative to fix the background colors.

制作山丘石头 Making the Rocky Hills

1

让我们先看一下伊瓜苏瀑布的真实石头。

For inspiration, let's look at an image of the stones at Iguazu Falls first.

2

把黑色纸铺在背景上,画出山丘和巨石的形状。后面的瀑布应该在中间偏右一点。

Place the backing paper on top of the background. Draw the hills and cliffs. The waterfall in the back should be in the middle a little toward right.

3

把瀑布部分如图剪出来。

Cut the waterfall parts out as shown.

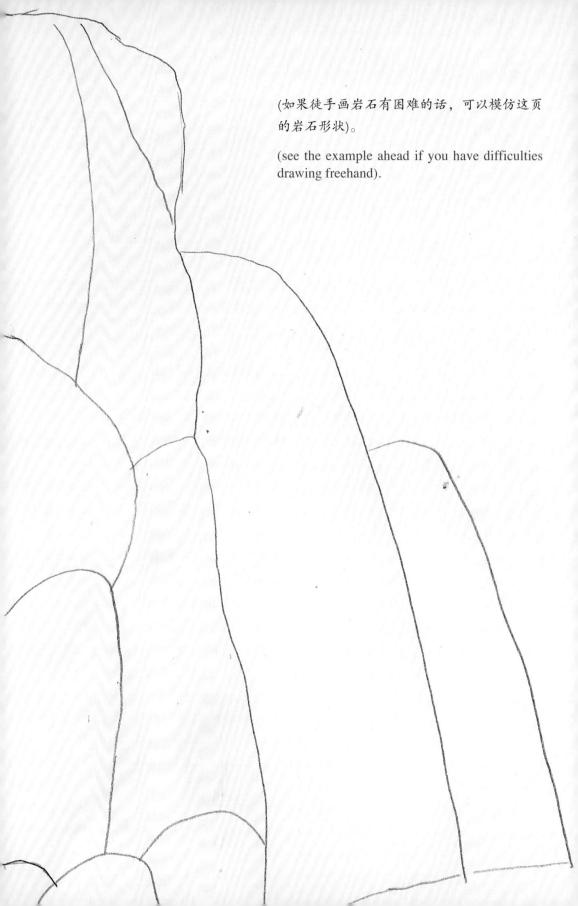

(如果徒手画岩石有困难的话，可以模仿这页的岩石形狀)。

(see the example ahead if you have difficulties drawing freehand).

风景画设计 Landscape Design

4

瀑布形状上贴满黑色叶子。

Glue black leaves on top of the waterfall shape to cover the entire shape.

5

把形状翻过来，用镊子夹断多余的。不要用剪刀剪。

Flip the shape over, try to break away the extras with tweezes. Do not cut with scissors.

6

需要边缘有右边这样不规则的锯齿状。
两片山丘形状都是同样制作。

It is good to have a jagged edge as show on the right.
Do the same for both cliffside shapes.

7

把巨石按照图样一一剪出来。

Cut large boulders into pieces according to the drawing.

8

用深色的叶子贴满靠近瀑布这片大的。右边边缘用镊子修成不规则锯齿状。左边可以不用修。

Glue some dark leaves on the piece next to the waterfall. The right side of the cliff needs to be created by using the jagged edge of the leaves. The left side can be left just the way it is.

用镊子把中间色和浅棕色叶子撕成条状。在深色大块岩石的顶部曲线边缘附近贴上中等棕色。然后是一个细的浅色条。这里有一个特写。

Use tweezers to break out strips of medium and light brown leaves. Glue the medium brown next to the top curved edge of the dark large cliffside piece. Then add a thin lighter strip. Here is a close look.

10

左上端的巨石也是同样做法。
底部岩石使用深绿色的叶子。

The big boulder on the top left is also done the same way.
 Use dark green leaves for the boulders below.

风景画设计 Landscape Design

11

最底部的两个大岩石可以用鳄梨壳完成。见下图。

Optionally, the very bottom two large boulders can be created with avocado shell. See pictures below.

12

底部用鳄梨壳做岩石。

Optional boulders created with avocado shell.

109

制作瀑布 Making the Waterfall

1

打开一个棉花球,拉出一条薄而长的条。然后再继续打薄。

Open a household cotton ball as shown. Pull a thin strip out. Thin the strip further.

2

顶端按照山丘顶的坡度折一下。

Fold the top to the same angle as the slope of the waterfall's edge.

3

用很少的胶稍微把顶端粘在山丘上。

Glue the cotton (go very light with glue) to the top of the waterfall.

风景画设计 Landscape Design

4

继续沿着山坡铺非常薄的一层棉花，形成瀑布形状。来自棉球的棉花含有类似水花的微小颗粒。

cotton from cotton balls contains tiny particles which resemble splashes.

5 从棉花中取出种子。去除坚硬的种子，留用其纤维。使用2~3粒种子的纤维形成长条状。

来自真实棉球的纤维更有韧性和弹性，与家用棉球相比，更容易拉成细长的条状。用手指搓一下。由于岩层的形状，瀑布形成一些水柱，我们要为瀑布描绘同样的力量。我们也可以使用美丽异木棉的毛代替棉花，它的好处是有着植物的天然光泽，棉花则没有。

也可以尝试把马利筋种子的毛和棉花混合，也可以有光泽。马利筋种子的毛单独则难成型。天然的光泽可以让观者在摆动头从不同角度观看画作的时候，感受到水的流动。

Remove the seeds from a real cotton ball. Get rid of the hard seed but use the fiber. Use fibers from two or three cotton seeds to form a long strip.

Cotton fibers from the natural seed is stronger and more elastic. It is easier to pull into thin long strips than from household cotton balls. Twist the strip with your fingers. Waterfall form columns of water due to water flowing over rock formations. We want to depict the same effect for our waterfall. Optionally, you can use the floss from silk floss tree instead of real cotton. The advantage with silk floss is that the water column will have a natural shine which is not present with cotton.

We can also combine cotton and milkweed seed hair to obtain a shine. Milkweed alone is not good for twisting. The natural shine will give viewer the feel of moving water when they view from the waterfall from different angles.

6

搓过的棉条或美丽异木棉条顶端应该比较尖，顺着山的顶端方向稍微倾斜一边。

The tip of the twisted cotton or floss should be somewhat pointy and slightly bent to follow the hill top.

7

记住自然形成的瀑布永远都不是一幅平顺的水帘。它一定是水花四溅。
把中间这个瀑布贴好（这是右边的瀑布）。

Remember that a waterfall from the nature is never an even sheet of water flowing down. It always splashes.

Glue the waterfall toward center of the picture (to the right of the main waterfall).

8

还有一种做法就是利用天然的竹纤维制作水柱。
可以把纤维拉成瀑布水柱形状。

Another option is to use natural bamboo fiber to form the water columns.

We can pull fibers into the shape of the water columns.

制作水雾 Making the Mist

典具纸或丝纸是用于制造由瀑布引起的水雾缭绕之环境。如果你没有典具纸或丝纸，作品将跳过描绘这个细节，但瀑布应该看起来还不错的。

Tengucho or Sanwa paper is perfect for making a misty atmosphere caused by the waterfalls. If you do not have either tengucho or sanwa paper, your picture will not depict this detail but the waterfall should still look fine.

离我们远的瀑布因为水汽围绕，看起来会更加柔和。因此我们把它放在典具纸或丝纸后面。

The further we view a waterfall, the more subdued the water vapor surrounding it are. Hence we place it behind the tengucho or sanwa paper.

下面这步适用于计划真空密封画作的人。如果不打算真空密封，请跳过此步骤。

The following step is for those who plan to vacuum seal the picture. If you do not plan to, skip.

1

将典具纸或丝纸揉成一团。揉两三次，以确保皱纹比较细。然后仔细打开纸张。用手压平纸张，但不要拉扯纸张。

Roll tengucho or sanwa paper into a ball. Twist a couple of times to make sure the wrinkles are fine. And then open the paper up carefully. Flatten the paper with your hands but do not stretch the paper too much.

2

完全覆盖右侧非常重要。如果纸张稍短，左侧露出些许也无妨。

对于那些不打算真空密封的人，用平整的典具纸或丝纸覆盖画面。

It is important to cover the right side completely. It is fine if the paper is a little short to cover the left side.

For those who do not plan to vacuum seal, cover the picture with flat tengucho or sanwa paper.

3

然后我们用之前同样的方法制作前面这片瀑布。

Next we make the waterfall in the front using the same method as the one in the back.

4

再贴上之前制作的石壁。

Glue the stone cliff wall we made earlier.

可选择地使用一小块透明亚克力压下瀑布，检查真空密封后的模样。如果水太薄，现在是时候修理了。如果它太厚，你没有看到水柱之间有任何暗区，请将它变薄。我们不要一整张棉花或破烂的棉絮。

Optionally, use a small piece of clear acrylic to press down the waterfall to preview how it will look after vacuum sealing. If the waterfall appears is too thin, adjust to create the desired effect. If it is so thick that you do not see any dark areas between water columns, thin it down. We do not want a sheet of cotton or tattered cotton.

在瀑布底部放一些松散的棉花，就像飞溅一样。使用透明亚克力检查厚度。

Place some loose cotton on the bottom of the waterfall to resemble splashes. Use the acrylic to check thickness.

制作水中的石头 Making the Rocks in the Water

1　拿一片黑叶。底部需要保持平整，因为它浸没在水中。在顶部贴上一片深橄榄绿叶，然后用一片黄叶盖住最尖端。不要用剪刀剪这些碎片，而是用镊子夹断。修剪底部。

热带溪流中的石头顶端有苔藓生长故是深橄榄色。湿的部分非常暗，几乎是黑色。光从顶部照射，因此尖端颜色较浅。

有些石头没有苔藓生长，因此这些石头是深褐色的。可以使用深棕色叶子代替深橄榄绿色。

Take a piece of dark leaf. The bottom needs to be flat since it is submerged in water. Glue a piece of dark olive green leaf on the top. Then cap the very top with a piece of yellow leaf. Do not cut these pieces but rather use tweezers to nip the leaves. Trim the very bottom.

The stones in tropical rivers are dark olive in color with moss growth. The wet portion is very dark, almost black. Light shines from above so the very top is lighter in color.

Some stones do not have moss growth. Hence those stone are dark brown in color. Instead of dark olive green, we can use dark brown leaves.

只有光线照射的石头才会产生光影效果。峡谷阴影区的石头都是黑暗的。因此，我们只需要在黄色区域制作一些具有明暗效果的石头。记住石头也不是都具有相同的形状和大小。　2

Only stones that are being illuminated will have light and shadow. The stones in the canyon's shadow area appear dark in color. Hence, we only need to make a few stones that we place in the yellow area with the light and shadow effect. Also stones do not all have the same shape and size. Make sure you keep this in mind when making stones.

风景画设计 Landscape Design

水中的石头会使水在其周围流动并产生小波浪。我们使用小条棉条（或是竹纤维或是美丽异木棉）放在石头的底部来描绘这一点。

3

Stones in water would cause the water to flow around it and creating small waves. We use small strip of cotton (or bamboo fiber or silk floss) to place on the bottom of the stone to depict this.

4 右侧的大型岩层由叶子和鳄梨壳组合而成，因此我们可以拥有不同的纹理和颜色。

The large rock formation on the right is made of combining leaves and avocado shell so we can have different texture and colors.

石头上的苔藓 Moss Growth on Stone

从翠云草或苔藓中分离出小片叶片，选择性地粘在岩石的裂缝上。并非所有裂缝都会有苔藓生长。

1

Break tiny pieces of blades from peacock fern or moss to glue on the crack of the rocks selectively. Not all cracks would have growth.

山谷边缘的树 Tree Branches on the Edge of Canyon

1

首先我们掀开典具纸或丝纸，贴上一些深色的铁线蕨。这些蕨叶看起来不好也没关系。我们只是需要一些剪影。

First we lift the tengucho or sanwa paper to glue some dark maidenhair fern. It does not matter how dead looking these are since we just need some silhouette.

2

覆盖回典具纸或丝纸，然后从右上角延伸到画中间贴铁线蕨。

Cover the tengucho or sanwa paper back and maidenhair fern on the top from the upper right corner extending to the middle of the picture.

风景画设计 Landscape Design

再贴上少量较浅的绿色铁线蕨在上面，描绘阳光透过树叶的样子。如果喜欢的话，添加一些小的白色或黄色花。这幅画的主角是瀑布。所以花不能太多。如果使用像雪柳这样的花，需要有正面和侧面花才能看起来自然。
左上角也贴上铁线蕨。

Glue lighter green maidenhair fern on the top to depict sunshine coming through the tree leaves. Add a few small white or yellow flowers if desired. This picture is to emphasis the waterfall so go very light with flowers. If we are using small flowers such as bridal wreath, try having both full faces and side views to look natural.

Glue maidenhair fern on the top left.

定稿细节调整 Final Adjustments

在真空密封之前，确保瀑布上没有微小的碎片。拍摄完成的作品在电脑屏幕上看通常可以帮助您觉察到要修改的地方。

Before vacuum sealing, make sure there are no tiny plant particles, especially on the waterfall. Take a picture of the finished work and look at it on a computer screen. This usually helps you to see areas where you want to make modifications.

将完成的作品放入装有干燥剂的大盒子中过夜，以确保在真空密封前作品完全干燥。

Place the finished work in a large box with active desiccant overnight to ensure the work is completely dry before vacuum sealing.

装裱注意 Note on Framing

这种设计的最佳装裱方法是真空密封。如果您没有计划真空密封，请确保图片和玻璃之间没有任何空间。

The best way to frame this design is vacuum sealing. Just in case you are not vacuum sealing, make sure the picture is pressed against glass. There should not be any space between the picture and the glass.

海浪
Wave

简介 Introduction

本课程细致地讲解这幅大师级压花画的技巧。我们学习海洋风景设计和海水的光影基础。

This class details the techniques used in this masterpiece of pressed flower art. We learn about seascape designs and some basics about light and shade on bodies of water.

材料 Materials

1. 8X10 英寸白色卡纸或水彩纸
2. 双面贴
3. 无酸玻璃胶
4. 猫或狗毛刷
5. 鼠标垫
6. 花材：
   a. 三种不同深浅蓝色绣球（浅色、中间调和深色）
   b. 白色绣球
   c. 白雪木
   d. 深蓝翠雀
   e. 蓝绿色老绣球
   f. 白千层
   g. 4~5 不同色调绿色叶子

风景画设计 Landscape Design

1. 8"×10" white card stock or watercolor paper
2. Double sided adhesive
3. Acid free silicone glue
4. Cat or Dog brush
5. Mouse pad
6. Pressed flowers:
   a. three different blue shades hydrangea (deep, mid tone, light)
   b. White hydrangea
   c. Diamond frost euphorbia
   d. Deep blue delphinium
   e. Green/blue aged hydrangea
   f. Paper bark tree bark
   g. 4-5 different shades of green leaves prefer having spots

海的色彩 Colors of The Sea

一天中的时间和天气状况对海洋色彩的影响很大。海水反射了天空中的光。在一天中的不同时间和不同的天气条件下，颜色看起来会有所不同。我的作品是关于南加州的海滩。我们这里以晴朗的天空、温暖的阳光而闻名。

Time of the day and weather condition play very important parts of the colors of the sea. The sea water reflects the light of the sky. Colors would look different with different time of the day and different weather conditions. My picture is about a Southern California beach. We are known for clear skies, warm and sunny weather.

远处的水是很深的蓝色。当我们从深海往陆地走时，水开始变成蓝绿色。离岸边越近，水的颜色越浅，而且变成绿色。当波浪升起时，如果螺旋状翻滚上升（图2），则顶部是较浅的颜色，几乎是透明的，因为光线正在穿过波浪。但是，如果波浪上升但仅膨胀（图1），则波浪包含大体积的水让光线无法穿透，因此它颜色很深。

The distant water is a very dark blue. As we come toward the shore, the water starts to turn to a teal color. Closer to the shore the water is

图 1　　　　　　图 2

shallower, and it turns green. When a wave rises up, if it spirals over (picture 2), the top is a lighter color almost transparent because the light is passing thru the wave. However, if the wave rises up but only swells up (picture 1), the wave comprises massive body of water , light cannot pass through so it is dark.

天空的色彩 Colors of The Sky

　　与大海相比，蓝色的天空颜色更浅。越接近地平线，它的颜色越浅。

The color of the sky is lighter in blue compare to the sea. It is even lighter in color as it is toward the horizon.

设计理念 Design Concept

　　这张画的设计使我们可以根据自己的喜好制作细节。例如，可以使用粉彩、水彩或丙烯染一个天空或用彩色纸做一个天空。另一个例子是，可以不做远距离海面上的波纹，而只是做一个填满蓝色花瓣的大海。

This picture is designed so that we can have as much details or as less as we want. For example, one can have a painted sky with soft pastel, watercolor or acrylic; or use color paper for the sky. Another example would be that we do not make the very fine details of distance waves but just leave the open sea with blue colored petals.

构图和底稿 Sketch and Drawing

我喜欢海滩，大海以及在加州海滩上度过的温暖夏日。

I love the beach, the sea, and the warm summer days spent on the beaches of California.

我先画了一幅海浪的简图。

I did an initial study of the wave.

然后根据这个简图，画了一个适合压花的底稿。

I then drew a picture for my pressed flower design according to the initial study.

我们可以把它用白色卡纸打印出来制作这幅画。

We can print out the picture on a piece of white card stock to make this picture.

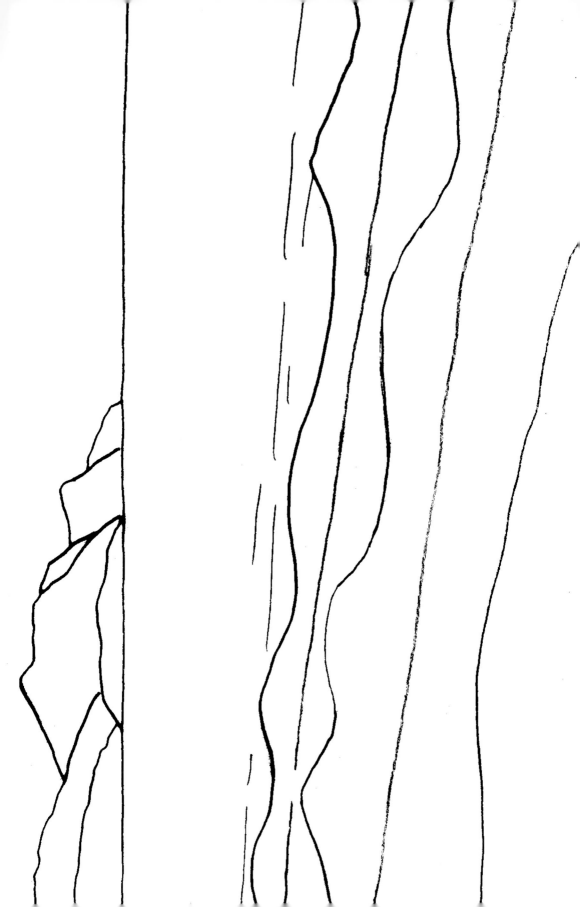

制作天空 Making the Sky

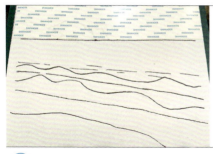

1. 将双面贴覆盖在天空部分（包括远山部分）。

Place double sided adhesive on the space for sky (including the distance mountains).

2. 使用镊子掰碎花瓣的边缘。或者，您可以使用一把波纹状的边缘剪刀造成花瓣边缘不平整。

Use tweezers to break the edges of the petal. Or you can use a pair of ripple or deckle edge scissors to cause uneven edges of the petal.

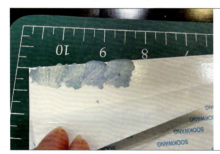

3. 把离型纸折下来，贴处理好边缘的浅蓝色绣球花瓣。掰碎边缘可以使花瓣更能融合成一体。

Fold the protective paper and glue light blue hydrangea petals. Torn edges would make the blending better.

4.

稍微压着山体边缘的线条。

Slightly overlap the edge of the mountain.

5. 用浅蓝色的绣球花瓣覆盖整个天空。确保保护纸仍覆盖在山脉上。

Cover the entire sky with light blue hydrangea petals. Make sure the protective paper is still covering the mountains.

制作山丘 Making the Mountains

使用不同的绿色阴影树叶覆盖山脉。使用镊子将碎片弄碎，以适应所需区域。根据图纸粘贴在该区域上。确保顶部略微与天空重叠，以免出现黑线。另外，请确保下方在水平线下延伸，以使黑线也被遮盖住。

Use different green shaded leaves to cover the mountains. Use tweezers to break the pieces to fit the areas. Stick onto the area according to the drawing. Make sure the top would slightly overlap the sky so there would be no black lines showing. Also, ensure the lower portion extends just below the horizon so the black line is also covered.

首先制作高山。如果在较高的山脉和较低的小山之间出现一些黑线，那也无妨。

Do the tall mountains first. It is fine if there are some black lines showing between the taller mountains and lower hills.

在较低的山丘上使用稍微亮一点的叶子。确保它们延伸到水平线以下覆盖黑线。

Use slightly brighter leaves for the lower hills. Make sure they extend slightly below the horizon covering the black line.

风景画设计　Landscape Design

制作大海色彩 Making the Color Base for the Sea

将一条细的双面胶带粘在一张白纸上。然后把纸裁至符合胶带的尺寸。

Glue a thin strip of double sided tape onto a piece of white paper. Cut the paper according to the size of tape.

把中等蓝色的绣球花瓣贴在双面贴上。确保没有白色的纸暴露出来。

Glue medium blue hydrangea petals onto the tape. Make sure there is no white shown on the paper tape.

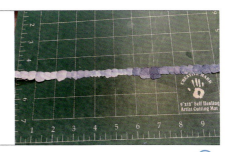

把带子翻转过来，把一端多余的花瓣按照直线的胶带修剪好。

Turn the tape over and trim one side of the hydrangeas to the straight edge of the tape.

把绣球花带子贴在画上形成地平线。

Glue the hydrangea tape onto the picture to form the horizon.

继续将中等蓝色的绣球花从绣球花带粘一直贴到波浪上方的线条，并稍微重叠线条。像上面一样准备另一个绣球带。将绣球胶带粘上，贴在波浪上方形成直线。

Continue gluing medium blue hydrangea from the tape to the top line above the wave and slightly overlapping the line. Prepare another hydrangea tape just like above for the horizon line. Glue the hydrangea tape to define the line above wave.

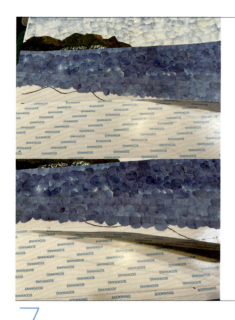

继续将中等蓝色的绣球花粘到两个波纹之间的线上。像上面一样准备另一个绣球带。将绣球胶带粘在线上,以定义两个波浪之间的线。

Continue gluing medium blue hydrangea to the line between two waves. Prepare another hydrangea tape just like above for the horizon line. Glue the hydrangea tape to define the line between two waves.

将中等蓝色的绣球粘贴至波浪下方的线。

Glue medium blue color hydrangea all the way to the line below the wave.

制作海滩 Make the Beach

剥白千层树皮,让它露出看起来像沙子的一面。树皮应由3~4层组成,足够稳定以支撑较大的尺寸。将其撕成大约海滩的形状,贴好。

Peel a piece of paper bark tree bark to reveal the side that looks like sand. The bark should consist of 3-4 layers to be stable enough to support the larger size. Shape it to about the beach shape. Glue down.

制作浅水 Make the Shallow Water

用蓝绿色老绣球填补空隙。

Glue green/blue hydrangea petals to fill the gaps.

设定海浪 Define Waves

将白色的绣球花放在鼠标垫的背面。用宠物刷敲打花朵,刺出很多小孔。用镊子把花瓣掰碎。对光拿起图。将白色小花瓣片粘到海浪的高低处。

Place a white hydrangea on the back of the mouse pad. Beat the flower with the cat/dog brush to punch a lot of tiny holes. Use tweezers to break small pieces of the petal. Lift the picture against light to see the lines. Glue the small white petal pieces to the highs and lows of the waves.

使用带有孔片的白色花瓣小片连接高点和低点以形成波浪。

Use the small white petal with holes pieces to connect the highs and lows to form waves.

制作涌浪 Making the Swell Wave

将更多的白色小片粘在画上形成飞溅物或泡沫。将深蓝色的小片贴在波浪的阴影处。另外，由于海浪成螺旋形，水会旋转并导致阴影形成线条状。

Glue more white pieces for the splash/foam. Glue dark blue pieces for the shade of the wave. Also, because the wave spirals, the water swirl and cause the shadow form lines.

制作翻滚浪 Making swirl wave

将蓝色或绿色的绣球花瓣剪成细条。我使用颜色较浅的一面。将它们倾斜粘在波浪的底部。把整个波浪贴满。

Cut the blue/green hydrangea petals into thin slides. I use the side with lighter color. Glue them diagonally onto the bottom of the wave. Fill the entire wave.

风景画设计 Landscape Design

找到一些薄而且最浅色的中等蓝色花朵，用宠物刷敲打花瓣，然后将其掰成小块，贴在绿色条上。这样可以使条状的地方看起来比较顺滑并能更好地融合色彩。贴一些深蓝色的花瓣碎片在底部形成阴影。

Find a few very thin and lightest medium blue flowers, beat the petals with pet brush and then break into small pieces to spot glue onto the green strips. This would smooth the harsh slides and blend colors better.Glue some dark blue petal pieces on the bottom of the wave for shadow.

增加几行浪 Add Rows of Waves

距离海滩较远的海浪不那么高。我们站在海滩上常常只看到一些白色的泡沫滚动。使用牙签把胶弄成一条线，然后将白色打细小孔的花瓣小片粘成一条线。

Waves further from the beach are not as high. We often just see some white foam rolling in standing on the beach. Use toothpick to make a glue line and then glue small pieces of white beaten petal pieces to form a line.

在白线下贴上深蓝色花瓣碎片为阴影。

Glue dark blue hydrangea petal pieces under the white line for shadows.

131

增强阴影 Shadow Strengthening

1

用宠物刷敲打深蓝色的翠雀花瓣。将胶水涂在整个花瓣上,但不要太多。

Beat the deep blue delphinium petals with the pet brush. Apply glue onto the entire petal but not too much.

2

将花瓣断成两半,然后将其粘在波浪的底部,以增强阴影。使用镊子的尖端进一步破碎翠雀花瓣。

Break the petal into two and then glue down to the bottom of the wave to strengthen the shadow shade. Use the tip of the tweezers to further break down the delphinium petal.

飞溅 Splash

使用白雪木的花头制作飞溅的浪花。在波浪的顶部使用一两个花头。首先为泼落的海浪下方粘上一行,然后错开在上面粘上另一行。用白色的小花瓣贴在白雪木绿色部分这样只有白色显露。

Use diamond frost euphorbia flower head as splashes. Use one or two flower heads on the top of the wave. Glue a row for the lower under spell first and then another row on the top. Glue small white petal pieces to block the green parts of the euphorbia.

泡沫 Foam

用宠物刷敲打白色的绣球花。将每片花瓣弄成两半。用镊子掰破边缘。沿沙子边缘(在树皮上)粘一两行。

Beat the white hydrangea flowers with pet brush. Break each petal into two. Break the edges with tweezers. Glue one or two rows of these along the edge of the sand (on top of bark).

涟漪 Ripples

将一张双面贴粘在白纸上。将白色绣球花瓣贴满所有空间。剪一些两头尖的长条。再剪一些左侧有一带钩的。将这些贴在海滩的白色泡沫和下面的海浪之间。

Place a piece of double sided adhesive onto white paper. Glue white hydrangea petals to fill all the spaces. Cut some long thin strips with pointy ends. Have a few with hook on the left side. Glue these on the space between the foam on the beach and the lower wave.

远处的波浪 Distance Waves

1　把双面胶粘到白纸上。将深蓝色的绣球胶粘到一张上。将蓝绿色绣球胶粘到另一张上。剪成枣核型小片。混合两种颜色的小片。注意透视：远处的海面我们需要较小的小块紧密拼贴。当我们接近海滩，小片慢慢变大，小片之间的距离也越来越大。

Please double-sided adhesive onto white paper. Glue deep blue hydrangea onto one. Glue blue/green hydrangea onto another. Cut tiny pieces with two pointy ends. Mix the deep blue and green/blue pieces. Pay attention to the perspective. We want smaller pieces and tightly packed for the distance sea. As we work toward the beach, the pieces are larger and so are the distance between pieces.

2　把波浪小片贴到画面中间的白色海浪那里。

Glue the wave pieces all the way to the white foams in the middle of the picture.

风景画设计 Landscape Design

后记
Postface

历时三年，重写数次之后，我终于把这套书写完了。

非常感谢中国林业出版社的编辑印芳一直以来的支持。

非常感谢我的家人给与我的鼓励和支持，感谢我儿子帮忙为我校正英文版本，感谢我先生给我写作方向的一些启发。

感谢国际压花协会的老师们和会员们给我的鼓励，并且在日常的交流中、各种活动中还有不同的课程中给我很多启发。

我希望这套书可以为每一位喜欢压花艺术的读者提供一些思路，从而创作出具有个人风格的艺术。也为大家介绍各种压花艺术独特的技巧和方法。

I finally finished writing this set of books after revising it for three years.

Thank you very much Yin Fang, editor of China Forestry Publishing House, for your continuous support.

I thank my family very much for the encouragement and support. Thanks to my son for helping me correct the English version. Thanks to my husband for giving me some inspiration on writing direction.

Thanks to the teachers and members of the World Wide Pressed Flower Guild for their encouragements and inspirations in daily communication, various activities, and in different classes. I hope my book will provide some ideas for every artist and to create art with their own personal style. I hope to also introduce some unique techniques and methods of pressed flower art.

2021.04